T0239383

Environmental Footprints and Eco-design of Products and Processes

Series Editor

Subramanian Senthilkannan Muthu, Head of Sustainability - SgT Group and API, Hong Kong, Kowloon, Hong Kong

Indexed by Scopus

This series aims to broadly cover all the aspects related to environmental assessment of products, development of environmental and ecological indicators and eco-design of various products and processes. Below are the areas fall under the aims and scope of this series, but not limited to: Environmental Life Cycle Assessment; Social Life Cycle Assessment; Organizational and Product Carbon Footprints; Ecological, Energy and Water Footprints; Life cycle costing; Environmental and sustainable indicators; Environmental impact assessment methods and tools; Eco-design (sustainable design) aspects and tools; Biodegradation studies; Recycling; Solid waste management; Environmental and social audits; Green Purchasing and tools; Product environmental footprints; Environmental management standards and regulations; Eco-labels; Green Claims and green washing; Assessment of sustainability aspects.

More information about this series at https://link.springer.com/bookseries/13340

Subramanian Senthilkannan Muthu
Editor

Environmental Footprints of Crops

 Springer

Editor
Subramanian Senthilkannan Muthu
SgT Group and API
Hong Kong, Kowloon, Hong Kong

ISSN 2345-7651 ISSN 2345-766X (electronic)
Environmental Footprints and Eco-design of Products and Processes
ISBN 978-981-19-0536-0 ISBN 978-981-19-0534-6 (eBook)
https://doi.org/10.1007/978-981-19-0534-6

© The Editor(s) (if applicable) and The Author(s), under exclusive license to Springer Nature
Singapore Pte Ltd. 2022
This work is subject to copyright. All rights are solely and exclusively licensed by the Publisher, whether
the whole or part of the material is concerned, specifically the rights of translation, reprinting, reuse
of illustrations, recitation, broadcasting, reproduction on microfilms or in any other physical way, and
transmission or information storage and retrieval, electronic adaptation, computer software, or by similar
or dissimilar methodology now known or hereafter developed.
The use of general descriptive names, registered names, trademarks, service marks, etc. in this publication
does not imply, even in the absence of a specific statement, that such names are exempt from the relevant
protective laws and regulations and therefore free for general use.
The publisher, the authors and the editors are safe to assume that the advice and information in this book
are believed to be true and accurate at the date of publication. Neither the publisher nor the authors or
the editors give a warranty, expressed or implied, with respect to the material contained herein or for any
errors or omissions that may have been made. The publisher remains neutral with regard to jurisdictional
claims in published maps and institutional affiliations.

This Springer imprint is published by the registered company Springer Nature Singapore Pte Ltd.
The registered company address is: 152 Beach Road, #21-01/04 Gateway East, Singapore 189721,
Singapore

Contents

About the Editor

Dr. Subramanian Senthilkannan Muthu currently works for SgT Group as the Head of Sustainability, and is based out of Hong Kong. He earned his Ph.D. from the Hong Kong Polytechnic University, and is a renowned expert in the areas of Environmental Sustainability in Textiles & Clothing Supply Chain, Product Life Cycle Assessment (LCA) and Product Carbon Footprint Assessment (PCF) in various industrial sectors. He has 5 years of industrial experience in textile manufacturing, research and development and textile testing and over a decade's of experience in LCA, carbon and ecological footprints assessment of various consumer products. He has published more than 100 research publications, written numerous book chapters and authored/edited over 100 books in the areas of Carbon Footprint, Recycling, Environmental Assessment and Environmental Sustainability.

Water Footprint of Fruits in Arid and Semi-arid Regions

Ommolbanin Bazrafshan, Hadi Ramezani Etedali,
and Zahra Gerkani Nezhad Moshizi

Abstract In this study, water footprint components including green water footprint, blue water, and gray water and its economic value in the production of dates, almonds, and walnuts in Iran were estimated. The results showed that in dates, almonds, and walnuts, the share of blue water was 76.6, 71, and 90%, respectively; The share of green water is 9.6, 19%, the share of gray water is 13.5, 10, and 6%. The economic value of water in these three products is $0.6, $0.2, and $0.9 per cubic meter, respectively. The total volume resulting from the production of these three products in Iran is 13830 MCM per year, of which more than 90% is exported to neighboring countries. Water resources are very limited in Iran, especially in the southern and middle regions with arid and extra-arid climate, which water scarcity is increasing, due to the result of the government's water policy over the past two decades. During the past decades, self-sufficiency in strategic products such as wheat was the Iran government's policy. There was no control over productions, developments, or stopping the cultivation, exporting and importing of agricultural products with a view of economic value of water footprint and water footprint. Therefore, this perspective should always be at the top of the planning and planting pattern.

Keywords Water footprint · Economic value · Blue water · Green water · Fruits · Arid and semi-arid · Almonds · Date palms · Walnuts

O. Bazrafshan (✉) · Z. G. N. Moshizi
Department of Natural Resources Engineering, Faculty of Natural Resources and Agricultural Engineering, University of Hormozgan, Bandar Abbas, Iran
e-mail: O.bazrafshan@hormozgan.ac.ir

H. R. Etedali
Department of Water Sciences and Engineering, Imam Khomeini International University (IKIU), Qazvin, Iran

© The Author(s), under exclusive license to Springer Nature Singapore Pte Ltd. 2022
S. S. Muthu (ed.), *Environmental Footprints of Crops*,
Environmental Footprints and Eco-design of Products and Processes,
https://doi.org/10.1007/978-981-19-0534-6_1

1 Introduction

Iran ranks first in fruit production in the Middle East and North Africa. In recent years, Iran has been ranked eighth among the top 10 fruit producing countries in the world. Iranian fruit products include Persian walnut, citrus, kiwi, dates, cherries, pomegranate, oranges, grapes and raisins, almonds, pistachios [1].

Jean [2] writes that in Iran there were the same kinds of fruits as in Europe, and much more, all delicious. He mentions the great variety of melons, cucumbers, grapes, apricots, pomegranates, apples, pears, oranges, berries, prunes, figs, pistachios, almonds, walnuts, hazelnuts, and olives [3]

Water resources are very limited in Iran, especially in the southern and middle regions with arid and extra-arid climate, which water scarcity is increasing, due to the result of the government's water policy over the past two decades. During the past decades, self-sufficiency in strategic products such as wheat was the Iran government's policy. There was no control over productions, developments, or stopping the cultivation, exporting and importing of agricultural products with a view of virtual water, economic water footprint and water footprint [4].

Due to the climatic, topographical and altitude difference of today's Iran, a variety of fruits can be found in this country, from subtropical dates to fruits growing in temperate and cold regions [5, 6].

According to the FAO [7], Iran is among the top 7 countries in the production of 22 important agricultural products. In fact, Iran ranks first in pistachio production, second in date production and fourth in apple production in the world.

On the other hand, according to the FAO, the country's most important products are wheat, rice, other cereals, sugar beets, fruits, nuts, cotton and tobacco. Iran also produces dairy products, wool and a large amount of wood. Interestingly, in this country, irrigation areas are fed by modern water storage systems or by the ancient system of aqueducts. Thus, Iran is ranked 13th among 14 countries in the Middle East and North Africa.

More than three-quarters of the planet and our bodies are made up of water. Water has been the cause of human civilization. And various ways of extracting water in Iran, including various types of dams, Qanats, irrigation culverts, and water use in other sectors such as hydrotherapy, energy production, agriculture, industry, and mining. More than 75% of Iran is composed of arid and semi-arid climate regions. Its average rainfall is one-quarter of the world average (250 mm) and the lack of proper spatial and temporal distribution of precipitation is necessary to study water in Iran. Therefore, due to the situation in Iran and population growth, any planning at the micro and macro levels without studying water resources is opposed to sustainable development [8].

The continuous circulation and distribution of water in the earth and the atmosphere is called the hydrological cycle. The hydrological cycle is a complex cycle that takes place in the atmosphere, hydrosphere, and lithosphere; The most important elements of the hydrological cycle are 1—Rainfall 2—Surface runoff 3—Evaporation from water or soil surface 4—Transpiration from vegetation surface 5—Infiltration and 6—Underground flows [9].

In the hydrological cycle, water resources are divided into two categories: green water and blue water.

Water that is stored as moisture in the soil is called green water. Green water was first introduced in 1990 by Malin Falcon Mark. Green water is one of the important sources of water supply for plants, especially in rainfed lands. The amount of green water that a plant consumes for growth and development and is stored in plant tissue or lost during the transpiration process is known as virtual green water. The share of green water in agricultural production is much higher than irrigated water. The production of about 85% of agricultural products is directly dependent on green water, and this shows the potential and importance of green water in providing food to the people of the world [10]. While different water sources such as groundwater, surface water, springs, rainwater, Qanats are called blue water. Blue water is one of the important sources of water supply for plants in irrigated lands.

Water footprint is an indicator to indicate the volume of water, which is used directly or indirectly to produce goods or provide any services which includes the total water consumed during the production chain of a product [11].

According to Van Oel et al. [12] the water footprint consists of two components, blue water, and green water. Blue water is the volume of freshwater that is used from global sources of water, for example, (rivers, groundwater) to produce goods and services. Green water is a global sources of green water, i.e., rainfall stored in the soil is used.

In the agricultural sector, gray water footprints are also calculated. Gray water is the volume of water that is contaminated during the production process of agricultural products and has lost its original quality [12]. Today, in addition to the water crisis in the world, the problem of water pollution has intensified. The agricultural sector is considered as the most important pollutant of water resources. Global estimates show that about 450 cubic kilometers of wastewater are discharged into freshwater resources each year [13]. To dilute this amount of wastewater and recycle it, we need 6,000 cubic kilometers of water, equivalent to 67% of the total annual fresh water on Earth [13].

Iran is under arid and semi-arid climate. Therefore, it is facing a shortage of water resources. In recent years, factors such as population growth and, consequently, urban development, as well as industrial development, have led to an increase in demand for water. Therefore, in the current situation, there is a serious need for proper management of existing water resources and planning for their allocation [14]. On the other hand, climate change and frequent droughts have exacerbated water shortages. Therefore, there is a need to manage water demand in the agricultural sector. One of these strategies is to use the concept of water footprint of agricultural products, virtual water, and trade based on it [15]

The Water footprint is defined as an index for the allocation of freshwater resources and is used to formulate strategies for the allocation of water resources in an area [16].

Therefore, in order to allocate water to the agricultural sector, special attention should be paid to the concept of water footprint (physical and economic) and productivity of agricultural and horticultural products. In this chapter, Initially, the required data was introduced in the data source and methodology section. In the methodology section, the equations related to the calculation of water footprint and the economic value of water footprint were presented. The results of water footprint and its economic value in dates, almonds, and walnuts are presented. In each agricultural product, changes in yield per hectare, cultivation area, annual production, effective rainfall, green, blue, and gray water footprint, virtual water volume, and economic value of water in Iran are presented and each product in terms of two indicators water footprint (WF) and water footprint economic value (WFEV) is compared with other important agricultural products in Iran. Finally, a conclusion and summary of the research was presented.

2   Data Source and Methodology

The data used in this research are divided into two categories: crop data and climatic data. The crop data includes cultivated area, yield per hectare, chemical fertilizer, irrigation efficiency, crop coefficient, planting calendar, and soil type, which are collected from the Ministry of Agriculture Jihad organization (MAJ). In addition to this information, the climatic data are gathered from Iran Meteorological Organization (IRIMO) for each region during study period. The climatic variables contain the average of 10 years precipitation (mm), relative humidity (%), sunshine hours (hr), maximum temperature (°C), minimum temperature (°C), and wind speed (km/day) for the given duration.

2.1   Calculation of Water footprint components

The water footprints are calculated by applying the Hoekstra and Chapagain's framework [17].

The water requirement, irrigation requirement, and effective precipitation are computed using the CROPWAT model. In this regard, the crop evapotranspiration and water requirement is calculated by FAO-Penman–Monteith under the standard and non-standard conditions using the CROPWAT model [18]:

$$WF_{Green} = \frac{(P_e) * 10}{Y} \tag{1}$$

$$WF_{Blue} = \frac{(ET_c - P_e) * 10}{Y} \tag{2}$$

$$WF_{Grey} = \frac{\alpha * NAR}{C_{Max} - C_{Nat}} * \frac{1}{Y} \tag{3}$$

where 10 is the conversion factor from mm to m^3/ha. The $WF_{Green}WF_{Green}$, $WF_{Blue}WF_{Blue}$, and $WF_{grey}WF_{Grey}$ are water footprints in units of m^3/kg [19]. The P_e(mm) represents the sum of effective rainfall during growing season, which can be obtained using the USDA S.C. Method [20], and Y crop yield in mature almonds (ton/ha). α is the percentage of nitrogen fertilizer loss (α is considered 10% under irrigation and 5% under rainfed conditions NAR is the rate of used nitrogen fertilizer (kg/ha), C_{Max} is the critical concentration of nitrogen fertilizer (kg/m^3). C_{Nat} is the real nitrogen concentration in the receiving water (kg/m^3) [20].

Through this study, the $WF_{Gray}WF_{Grey}$ is computed only for nitrogen fertilizer where the maximum nitrogen concentration based on the US-EPA in receiving standard water is 10 mg/l. In addition, the C_{Nat} is considered as zero due to the lack of information about the real nitrogen concentration in receiving water [20]. Finally, the national WF components were estimated by taking the average of each component over all the provinces weighted by the share of each province in the whole almond production of the combined provinces [21].

Once the provincial WF components are calculated for each selected province, the total volume of WF components in each province and national scale can be obtained as the weighted average of WFs of crops as shown in Eqs. (4) and (5).

$$WFV_x = \sum_i WF_{i,x} \, Prod_{i,x} \quad i = 1, \ldots, 29, \; x = blue, \; green, \; gray \tag{4}$$

$$AWF = \frac{\sum_i WFV_{i,x}}{\sum_i Pr \, od_{i,x}} \tag{5}$$

where i is the index of the province, x is WF components (blue, green, gray), $Prod_{i,x}$ is the amount of almonds that are produced in the ith province under rainfed and irrigated conditions (kg), WFV_x is the total volume of each WF component (MCM), AWF is the national weighted average of each WF component under irrigated and rainfed conditions (m^3/kg).

2.2 Economical Value of Water Footprint

The economic value of WF components can be given as

$$WF_{EV} = \frac{NB}{WF_{Green}} \tag{6}$$

$$WF_{EV(\text{Blue})} = \frac{NB}{WF_{\text{Blue}}} \tag{7}$$

$$WF_{EV(\text{Grey})} = \frac{NB}{WF_{\text{Grey}}} \tag{8}$$

The $WFEV_{(\text{Green})}$, $WFEV_{(\text{Blue})}$, $WFEV_{(\text{Gray})}$ are the economic value of WF components (USD/m^3) and NB is the net benefit (USD/kg).

3   Water footprint components in date palm

3.1   *Introduction*

According to the World Food Organization, there are currently more than one billion and 353 thousand hectares of groves in the world, which annually about eight billion and 460 thousand tons of dates are harvested from these groves and sent to various markets, while the average date production in the world 6,252 kg per hectare [7].

Date palms are a fruit with high nutritional value due to their significant sugar content (about 70%). In addition to fresh consumption, it has many uses in industry. Also, this fruit is of special importance in Iran in terms of job creation, economic income generation, and especially the possibility of export and currency exchange [22].

The productive trees of the date palm in Iran are covered by 199,114 hectares, which they yearly produce about 1,014,006 tons of date palm. The main provinces of date palm producer are Kerman, Khoozestan, Sistan & Baloochestan, Hormozgan, Fars, South of Keram, and Bushehr (Fig. 3). These provinces contain 90% of culti-vated areas which produces 93% of date palm productions. According to MAJ [23], 92% of date palm orchards are irrigated lands in Iran.

There are several types of cultivar date palm in Iran, including soft, semi-dry, and dry. The soft date palm includes Mazafati, Bazmani, Parkoo, Abdollahi, Shahani, Kabkab, Rabbi, Khass, Breim, Gantar, Shekari, Kalooteh, Halavi, Mordasang, Khanizi, Khassoei, AAl-Mehtari, Mosalla, Mosalla, Haloo, Bardian, and Zardak. The semi-dry date palm contains Zahedi, Estamaran, Khazravi, Nikdini, and Halileh. Further, the dry date palm contains Piarom, Daski, Barhi, and Deiri [24].

3.2   *Cultivation Area, Crop Yield, Production, and Chemical Fertilizer Consumption*

Cultivated area (Fig. 2a), crop yield (Fig. 2b), production (Fig. 2c), and chemical fertilizer consumption (Fig. 2d) in each cultivar are represented in Fig. 2. The share of these variables is presented in Fig. 3. The most cultivated areas (78.2%) and date palm production (68.3%) in Iran belong to the soft type (Fig. 3). In this regard Sistan & Baloochestan province has the highest date palm production and the south of Kerman contains the highest cultivated area. The average production of soft date palm in Iran is about 5 tons/ha, which on average consumes 176 kg/h chemical fertilizer annually.

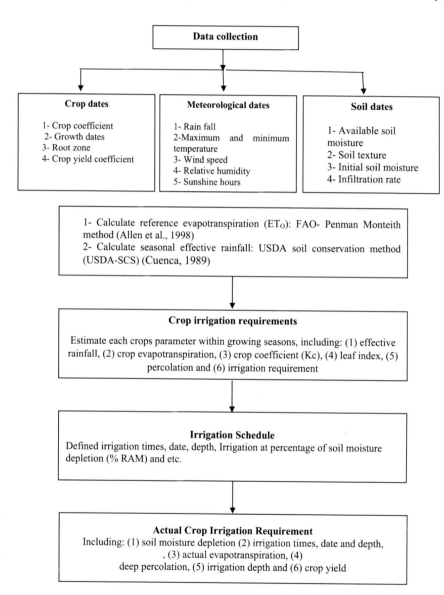

Fig. 1 Flow chart of CROPWAT irrigation management model

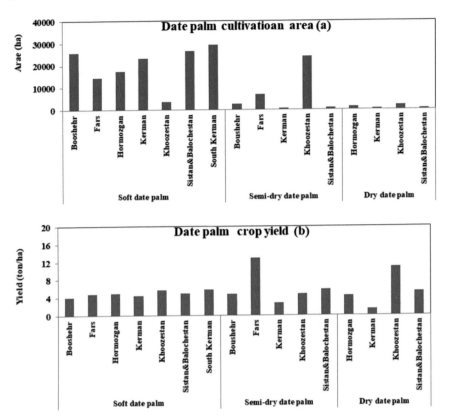

Fig. 2 Cultivation area (**a**), crop yield (**b**), production (**c**), and chemical fertilizer (**d**) each province for various date palms in Iran

The second rank of production and cultivation areas belongs to the semi-dry date palm with 27.9 and 19.4%, respectively. Five of those provinces produce the semi-dry date palm with 6.2 tons/ha crop yield.

The lowest cultivated area (2.4%) and production (3.8%) belongs to dry date palm, which is cultivated only in four provinces. The average crop yield of dry one is 5.5 tons/ha, and Khoozestan has the highest cultivated area and production in semi-dry and dry date palm.

3.3 Water Footprint (WF) and Water Footprint Economic Value (WFEV)

The WF and WFEV's shares of various date palms are shown in Fig. 4. The average WF of the soft date palm (Fig. 4a) is 2.68 m³/kg, with the shares of 9.9% (green),

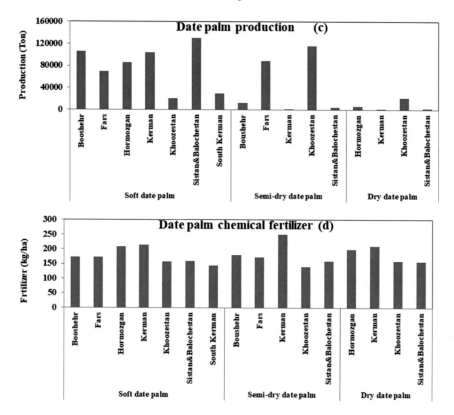

Fig. 2 (continued)

76.6% (blue), and 13.5% (gray) (Fig. 5a). The highest and the lowest of WF belongs to the Booshehr (3.42 m^3/kg) and Khoozestan (2.06 m^3/kg).

The average WFEV of the soft date palm is 0.49 USD/m^3 (Fig. 4a). This means that the value of 1 m^3 of consuming water to produce one kg of the soft date palm is about 0.49 USD. Also, Khoozestan (0.76 USD/m^3) and Booshehr (0.31 USD/m^3) have the highest and the lowest WFEV of the soft date palm in Iran.

The estimated WF of the semi-dry date palm is 2.56 m^3/kg (Fig. 4b). From this, the shares of green, blue, and gray WF are 10.1, 75.9, and 14% respectively. Further, the highest and the lowest WF belongs to Kerman (4.04 m^3/kg) and Fars (0.97 m^3/kg). The average WFEV of the semi-dry date palm is equal to that in soft date palm with 0.49 USD/m^3.

The dry date palm has the average WF of 3.83 m^3/kg, from which Kerman and Khoozestan contain the highest and the lowest WF with the values 8.06 and 1.08 m^3/kg, respectively (Fig. 4c). Moreover, the average of WFEV is 0.91 USD/m^3 that the lowest and the highest WFEV belongs to Khoozestan (1.46 USD/m^3) and Sistan & Baloochestan (0.26 USD/m^3).

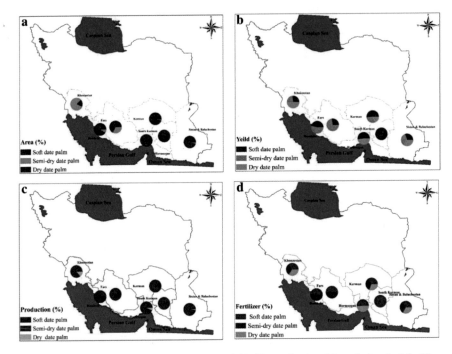

Fig. 3 The share of cultivation area (**a**), crop yield (**b**), production (**c**), and chemical fertilizer (**d**) each province for various date palms in Iran

Overall, the average WF of date palms in Iran is 3.02 m³/kg, with WFEV of 0.59 USD/m³. In this regard, the share of green, blue, and gray WF are 9.1, 76.9, and 12%, respectively. Furthermore, the highest and the lowest WF belong to the dry and semi-dry palms (Fig. 5).

3.4 Volume of WFCs in Each Cultivars

Figure 6 presented the volume of WFCs in different cultivars of date palm based on Eqs. 4 and 5. The volume of WFCs is calculated based on results, the total AWF of date pulms in Iran is 2347 MCM which the soft date cultivar with 1868.1 MCM has the largest share of water footprint in Iran. The share of volume of green, blue, and gray in soft date is 10, 78, 13%, respectively (Fig. 7a).

The total volume of WF in semi-dry and dry date palm cultivars is 423.1 and 55.8 MCM. The share of these is 21% from Total WF (Fig. 7b–d).

Finally, total AWF of date palms in Iran is 2347MCM which the share of blue WF is the largest share (1803.6 MCM). Also, considering the volume of gray footprints (311.3 MCM), the use of surface water and groundwater in arid and extra-arid regions such as southern Iran is not negligible (Fig. 6).

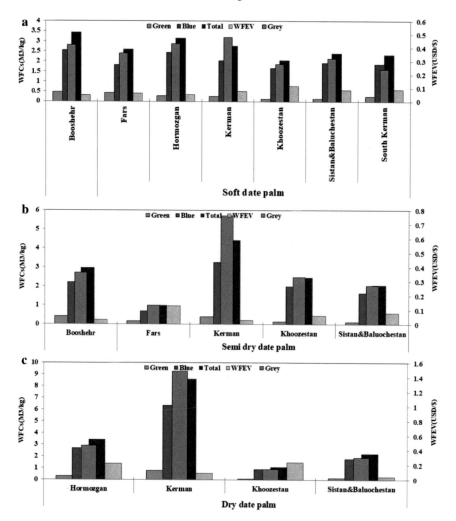

Fig. 4 Water footprint (WF) and water footprint economic value (WFEV) for soft (**a**), semi-dry (**b**), dry (**c**) in each province

4 Water Footprint in Almond

4.1 Introduction

Almond (Prunus amygdalus) is a fruit, the tree of which is native to Iran and the surrounding countries and Central Asia, but is cultivated in many areas. Almonds are one of the nuts that are well known for their unique properties [25].

Iran is one of the major producers and exporters of almonds. According to statistics and information from the Ministry of Jihad for Agriculture and the World Food

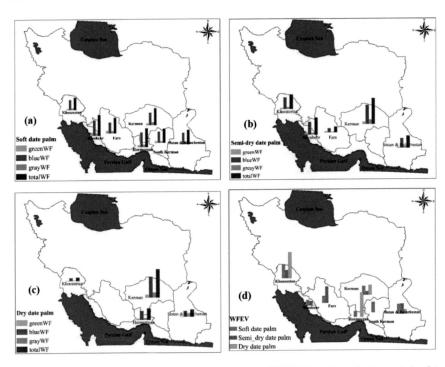

Fig. 5 Spatial distribution of water footprint components (WFCs) share for soft (**a**), semi-dry (**b**), dry (**c**), and water footprint economic value (WFEV) (**d**)

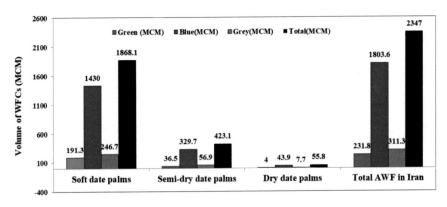

Fig. 6 Volume of WF components for date palms in Iran

Organization, with an area under cultivation of more than 140,000 hectares during 2004–2009, it is ranked third to fourth in terms of almond cultivation area, after Spain (about 6,501,000 hectares), the United States (about 290,000 hectares) and Tunisia (about 190,000 hectares). In terms of production, the situation in Iran is more favorable and during the studied years, it has been able to increase its ranking

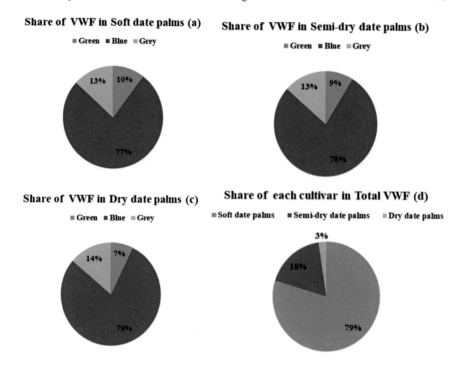

Fig. 7 The share of volume of WF components for soft (**a**), semi-dry (**b**), dry (**c**), and total (**d**) in date palms in Iran

from fifth to third place so that the amount of almond production has reached about 128 thousand tons and in the export sector has been able to be among the top ten exporting countries [26].

Water resources are inherently limited in these areas. More than 90% of the water resources used for irrigation are groundwater sources. So that more than 388 aquifers in Iran are in a critical situation and the production of many agricultural and horticultural products is facing severe challenges [14].

Twenty-nine provinces in Iran produce the almond. The cultivated regions are everywhere in Iran except the north, including the semi-arid, arid, and Mediterranean climate zone.

4.2 The Cultivation Area, Crop Yield, Production, and Chemical Fertilizer of Almond Production

The average of cultivation area (Fig. 8a), crop yield (Fig. 8b), production (Fig. 8c), and chemical fertilizer (Fig. 8d) in almond orchards are displayed in Fig. 8. The average cultivated almond area is 75,048 ha, with production of 130,000 tons per year. More

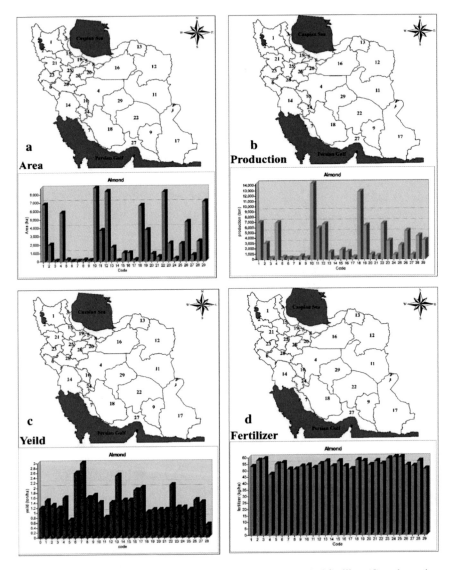

Fig. 8 Cultivation area (**a**), crop yield (**b**), production (**c**), and chemical fertilizer (**d**) each province for almond production in Iran

than 75% of these have been harvested from irrigated orchards [27]. The name of each province is presented in Table 1. Char Mahal Bakhtiari, Khorasan Razavi, Kerman, and Yazd have the greatest share in almond orchards (40%). However, based on results, the crop yield of Tehran (3 ton/ha) has the first rank in other provinces. In this regard, the yearly averages of nitrate chemical fertilizer consumption are about 54 kg/ha.

Table 1 Name and code of almond producing provinces in Iran

Code	Name of province	Code	Name of province	Code	Name of province
1	Azerbaijan Shargi	11	Khorasan Jonoobi	21	Kurdistan
2	Azerbaijan Gharbi	12	Khorasan Razavi	22	Kerman
3	Ardabil	13	Khorasan Shomali	23	Kermanshah
4	Isfahan	14	Khozestan	24	Kohgiluyeh Boyer-Ahmad
5	Alborz	15	Zanjan	25	Lorestan
6	Ilam	16	Semnan	26	Markazi
7	Boshehr	17	Sistan & Balochestan	27	Hormozgan
8	Tehran	18	Fars	28	Hamedan
9	South Kerman	19	Qazvin	29	Yazd
10	Chaharmahal & Bakhtiari	20	Qom		

4.3 Almond Water Footprint in Iran

The value of WF components and shares including green, blue, and gray for almonds is displayed in Figs. 9 and 10. The values of WF_{Green} are between 0.4 and 7.2 m³/kg, the amounts of WF_{Blue} are from 2.5 to 26.4 m³/kg, and finally the WF_{Grey} values ranged from 0.1 to 6.4 m³/kg. The average of WF_{green} is about 2.2 m³/kg in which the largest shares belong to Kohgiluyeh and Boyer-Ahmad (36%), Ilam (35%), and Kermanshah (34%), and the lowest shares obtained by Yazd (4%), Qom (6%), and Sistan & Balochestan (8%), respectively.

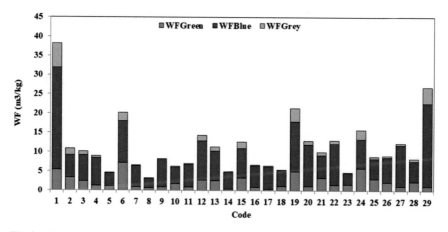

Fig. 9 WF components in almond at each province

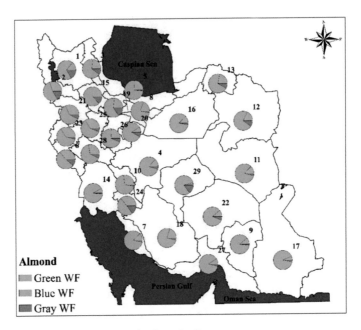

Fig. 10 The share of WF components in almond at Iran

The blue WF, with 8 m³/kg, contributes the largest share of WF in almond production. The results show that Sistan & Balochestan (90%), Khouzestan (89%), and Hormozgan (88%) have the largest shares of WF_{Blue}. In contrast, Kohgiluyeh and Boyer-Ahmad (49%), Azerbaijan Gharbi (49%), Kermanshah (54%), and Lorestan (60%) have the lowest shares in WF_{Blue} In this situation, the gray WF is calculated 1.1 m³/kg for almond production. According to the results, Azerbaijan Gharbi (16.8%), Qazvin (15.6%), and Azerbaijan Sharghi (15.2%) contribute the largest share of gray WF, while Khouzestan (2%) and Khorasan Jonoobi (2.4%) have the lowest shares.

4.4 *Economic Values of Water Footprint (WF$_{EV}$) in Almond*

Based on results, the average economic water footprint in almonds, in Iran is 5.16 m³/USD $. This means that for every 1 dollar of net income, 5.16 m³ water is used.

Based on WF$_{EV}$ values in Fig. 11, Azerbaijan Shargi (17.36 m³/US$) has the highest and Tehran (1.45 m³/US$) the lowest WF$_{EV}$ of almonds. Thus, Tehran has the lowest WF$_{EV}$ in almond production and provides higher economic value and lower water consumption to domestic and foreign markets.

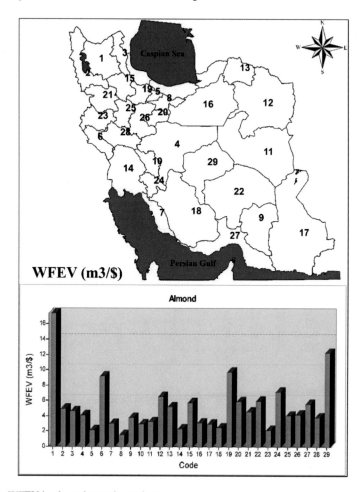

Fig. 11 WFEV in almond at each province

4.5 Volumes of Water Footprint Components in Almond Production

Figure 12 showed the share of green, blue, and gray WF in each province. The largest share of water consumption is related to water footprint and this situation exists in all provinces. The average share of blue, green, and gray water is 71, 18, and 10%, respectively. Figure 13 presented the volume of AWF in main provinces. AWF calculated based on Eqs. 4, 5. The highest volume of AWF belongs to Azerbaijan Sharghi (1849 MCM), Qazvin (943.8 MCM), and Khorasan Razavi (658 MCM). The average of AWF in almonds is 8256 MCM per year, in which the share of green, blue, and gray WF are 1493.7, 5888.9, 837.7 MCM, respectively (Fig. 14).

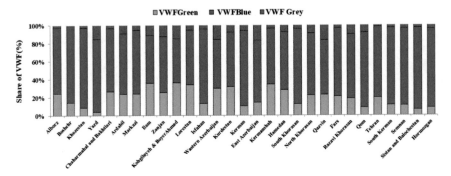

Fig. 12 The share of volume WFCs in subnational scale in almond production

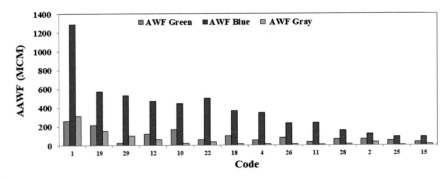

Fig. 13 Volume WFCs in subnational scale in 11provinces in almond production

Fig. 14 Total volume of
AWF in almonds at Iran

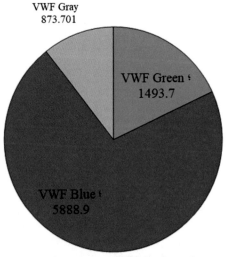

Average total WF in almond

5 Water Footprint in Walnuts

5.1 Introduction

Walnut tree (scientific name: Juglans spp.) is cultivated in China, Japan, France, and the United States. The most important walnut in the world is known as Iranian walnut, because it was first taken from Iran to the Middle East, and from there to Greece and Rome, then to England, and then to the United States. Persian walnut is the only species of this family whose kernels are economically consumed (Vahdati 2000). According to the World Food and Agriculture Organization (FAO), in 2017, the area cultivated world walnut cultivation was 1,097,700 hectares, which is an increase of about 68% compared to 2005. Among the countries of the world, in 2017, China with the production of 1,925,403 tons per year is in the first place, followed by the United States with 571,526 tons in the second place and Iran with 349,192 tons is in the third place. According to the latest information from Iran in 2014, it produced 9.11% of the world's walnuts [24].

5.2 The Sown Area, Total Production, and Yield of Walnut Production

The average yield (kg/ha) and cultivated area (ha), production (Ton), and chemical fertilizer (kg/ha) in walnut orchards are displayed in Fig. 15.

The average cultivated almond area is 110920 ha, with production 244,601 tons per year. More than 90% of these have been harvested from irrigated orchards [27]. Hamedan, Fars, Azarbaijan Sharghi, Kerman, and Kermanshah have the greatest share in walnut orchards (46%). The average crop yield of walnut in Iran is 2.2 tons/ha. However, based on results, the crop yield of Tehran (3.45 ton/ha) has the first rank in other provinces. In this regard, the yearly averages of nitrate chemical fertilizer consumption are about 68 kg/ha.

5.3 Walnut Water Footprint in Iran

The value of WF components obtained in walnut is displayed in Fig. 16. Also, the percentage of shares of each component at the provincial scale is presented in Fig. 16.

WF_{Green} varies in the range of 0–1.12 m^3/kg, WF_{Blue} 0.98–13.39 and WF_{Grey} 0.23–1.42 m^3/kg. The average total water footprint in walnut production on a national scale is 6.41 m^3/kg, of which 3% is green water, 90% is blue water, and 7% is gray water.

The average green water footprint is 0.2 m^3/kg, of which Ardabil, Mazandaran, and Guilan provinces have the highest share of 34.5, 15.6, and 7.6%, respectively, and Semnan and Chaharmahal & Bakhtiari provinces without Green footprint share (zero

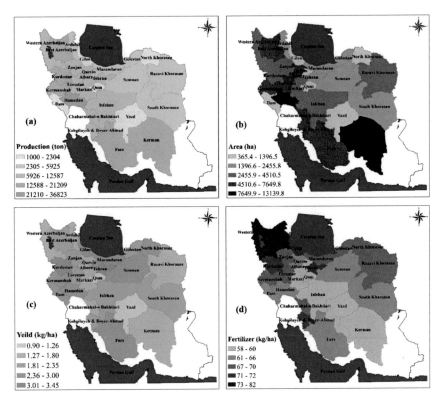

Fig. 15 Spatial distribution of production (**a**), sown area (**b**), crop yield (**c**), and chemical fertilizer (**d**) of walnut in Iran

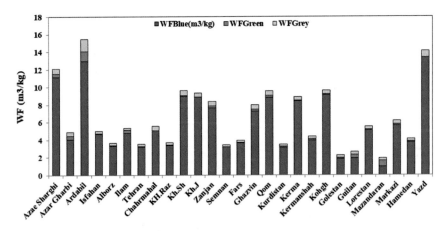

Fig. 16 WF component in walnut provincial producing

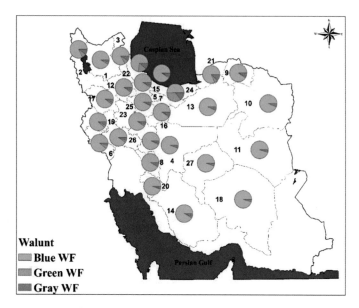

Fig. 17 The share of WF components in walnut at Iran

percent) have the lowest share of green water footprint. The average water footprint is 5.85 m³/kg. From this point of view, the provinces of Yazd, South Khorasan, Kohgiluyeh, and Boyer-Ahmad have the largest share of water with 95% of the total water footprint in the country. In front of Mazandaran provinces; Guilan and Golestan with about 50, 72.5, and 81.5% have the lowest share of water footprint. The average gray water footprint in Iran is about 0.41 m³/kg, with Mazandaran, Guilan, and Golestan provinces having the highest share of gray water footprint and Khorasan Jonoobi province with 14.9, 12.4, and 11.8%, respectively. Kerman, Kohgiluyeh & Boyer-Ahmad have the lowest share of gray water footprint with 4%.

5.4 Economic values of water footprint (WF_{EV}) in walnut

Figure 18 shows the spatial distribution of the economic value of walnut water footprints in Iran. According to the results, Mazandaran (2.97 USD/m³), Golestan (2.64 USD/m³), and Guilan (2.31 USD/m³) have the highest economic value of water footprint. Yazd province 0.41 USD/m³) has the lowest economic value of water footprint. After Yazd, East Azerbaijan, and Ardabil have the lowest economic value. In terms of spatial variation, it can be said that the central parts and parts of eastern Iran have the lowest and the northern parts of the country have the highest economic value of walnut water footprints.

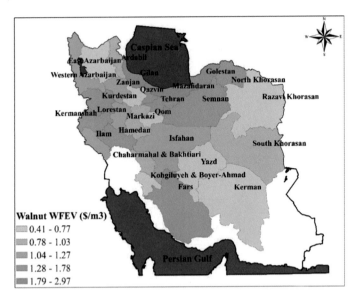

Fig. 18 Spatial distribution of WFEV in Iran for walnut production

5.5 Volumes of Water Footprint Components in Walnut Production

The results of the average volume of water footprint components on a provincial scale are presented in Fig. 19. The results show that on average, 3227 MCM of virtual water

Fig. 19 Total volume of AWF in walnut at Iran

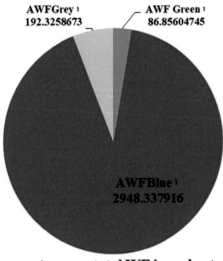

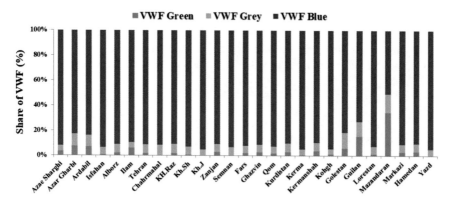

Fig. 20 The share of volume of WF in producing provinces of walnut

is consumed annually, of which 86 MCM are green water, 2948MCM are blue water, and 192 MCM are gray water. The share of each water footprint component at the provincial level is also shown in Fig. 20. In all provinces, the largest volume of virtual water is related to water. According to the above figure, in Mazandaran, Gilan, and Golestan provinces, the share of green water is 34, 15, and 6%, respectively. These provinces have a high potential for the development of rainfed cultivation. In fact, these three provinces are the only owners of rainfed walnuts in the country.

The share of blue water footprint in Iranian walnut is 90%, the resulting volume is 2945 MCM. More than 70% of the area under walnut cultivation in Iran is in arid and semi-arid provinces, which are faced with low effective rainfall, high evapotranspiration, and crop water requirements. More than 70% of the volume of water; it belongs to the provinces of Azerbaijan Sharghi (1493 MCM), Ardabil (202), Azarbaijan Gharbi (202 MCM), Isfahan (164 MCM), Alborz (150 MCM) (Fig. 21). On the other hand, in these provinces, irrigation losses due to improper irrigation methods have also increased the share of irrigation water footprint, so that lack of proper

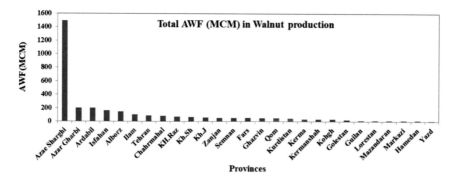

Fig. 21 Volume WFCs in provincial scale in walnut production

and improper management of irrigation methods, and the use of traditional irrigation methods has increased water footprint that the need to control irrigation losses, Increasing irrigation efficiency with the help of modern methods and proper irrigation history in these provinces seems to be necessary to increase water productivity and yield.

6 Conclusion and Summary

Water footprint and economic water footprint are appropriate tools for regional prioritizing in arid and semi-arid climates for all crops. Combining of water footprint and water footprint economic value with different scales provides appropriate information for managing water resources.

The purpose of current study is to analyze the water footprint and water footprint economic value in three Iranian export fruits including dates, almonds, and walnuts on a national and provincial scale. Iran ranks second in terms of date production, almonds in fourth place, and walnuts in third place. In the south of Iran, seven provinces produce dates with arid and semi-arid climates. Meanwhile, 29 provinces produce walnuts in almonds and walnuts. Figure 22 examines 35 Iranian agricultural products in terms of WF and WFEV. In this figure, WF = 1 belongs to a product that has the lowest water footprint and WF = 35 has the highest footprint among 35 products. However, WFEV = 1 belongs to the product with the highest economic value and WFEV = 35 has the lowest economic value. Comparison of these three fruits with 35 main agricultural products of Iran (Fig. 22), dates with = 23, WFEV = 20 WF; almonds are WF = 29, WFEV = 15 and walnuts with WF = 27, WFEV = 1. Also based on [24] min agricultural products at Iran in terms of WFEV are divided into six

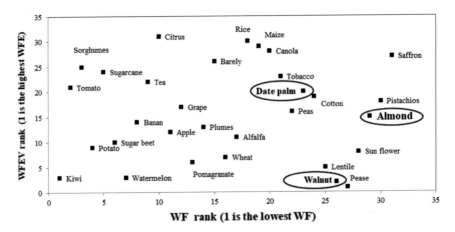

Fig. 22 WF and WFEV rank for date palm, almond, and walnut with main agricultural productions in Iran

categories: very low (0.01–0.49 USD/m^3), low (0.5–0.99), medium (1–1.89), high (1.9–2.49), very high (2.5–3.99), and excellent (>4). In this classification, almonds were in the middle category and dates and walnuts were in the very high category.

The total volume of water footprint of the production of these three fruits in Iran is equal to 13830 MCM. Of this amount, more than 90% is exported from Iran to neighboring countries such as the UAE, Turkey, Iraq, Syria, Russia, Qatar, and Lebanon.

Water resources are very limited in Iran, especially in the southern and middle regions with arid and extra-arid climate, in which water scarcity is increasing, due to the result of the government's water policy over the past two decades. During the past decades, self-sufficiency in strategic products such as wheat was the Iran government's policy. There was no control over productions, developments, or stopping the cultivation, exporting and importing of agricultural products with a view of virtual water, economic water footprint, and water footprint. Therefore, in virtual water export, the economic value of the exported water should always be at the top of the planning and planting pattern.

References

1. Khoshbakht K (2006) Agrobiodiversity of plant genetic resources in Savadkouh, Iran, with emphasis on plant uses and socioeconomic aspects. kassel university press GmbH
2. Chardin J (1735) Voyages en Perse et autres lieux de l'Orient (vol 3)
3. Matthee R (2009) The Safavids under western eyes: seventeenth-century european travelers to Iran. J Early Mod Hist 13(2):137–171
4. Abbasi F, Heydari N, Sohrab F (2006) Water use efficiency in Iran Islamic Republic: status, challenges and opportunities. AARINENA Water Use Efficiency Network, 58–70
5. Koocheki A, Ghorbani R (2005) Traditional agriculture in Iran and development challenges for organic agriculture. Int J Biodivers Sci Manag 1(1):52–57
6. McLachlan K (1988) The neglected garden: the politics and ecology of agriculture in Iran. IB Tauris and Co.
7. FAO (2020) FAOSTAT. http://www.fao.org/faostat/en/#country
8. Ardakanian R (2005) Overview of water management in Iran. In: Water conservation, reuse, and recycling: Proceeding of an Iranian-American workshop. The National Academies Press, Washington, DC, pp 18–33
9. Bierkens MF, Dolman AJ, Troch PA (eds) (2008) Climate and the hydrological cycle. International Association of Hydrological Sciences
10. Zygmunt J (2007) Hidden waters. Waterwise, London
11. Hoekstra AY, Hung PQ (2002) Virtual water trade: a quantification of virtual water flows between nations in relation to international crop trade. Value if the Watre Research Report Series. No. 11, UNESCO-IHE, Delft. https://thewaternetwork.com/_/integrated-water-res ource-management-iwrm/document-XAs/a-quantification-of-virtual-water-flows-between-nations-in-relation-to-international-crop-trade-TPANWVVuKrHT0upeMoSA1A
12. Van Oel PR, Mekonnen M, Hoekstra AY (2008) The external water footprint of the Netherlands: quantification and impact assessment
13. Mekonnen MM, Hoekstra AY (2011) The green, blue and grey water footprint of crops and derived crop products. Hydrol Earth Syst Sci 15(5):1577–1600
14. Madani K (2014) Water management in Iran: what is causing the looming crisis? J Environ Stud Sci 4(4):315–328. https://link.springer.com/article/10.1007%2Fs13412-014-0182-z

15. Lenzen M, Moran D, Bhaduri A, Kanemoto K, Bekchanov M, Geschke A, Foran B (2013) International trade of scarce water. Ecol Econ 94:78–85
16. Novoa V, Ahumada-Rudolph R, Rojas O, Sáez K, de la Barrera F, Arumí JL (2019) Understanding agricultural water footprint variability to improve water management in Chile. Sci Total Environ 670:188–199
17. Hoekstra AY, Chapagain AK (2008) Globalization of water: sharing the planets freshwater resources. Blakwell Publishing, Oxford, UK. https://doi.org/10.1002/9780470696224. http://onlinelibrary.wiley.com/book/10.1002/9780470696224
18. Allen RG, Pereira LS, Raes D, Smith M (1998) Crop evapotranspiration: guidelines for computing crop water requirements. FAO Drainage and Irrigation Paper 56, Food and Agriculture Organization, Rome
19. Bazrafshan O, Zamani H, Etedali HR, Dehghanpir S (2019) Assessment of citrus water footprint components and impact of climatic and non-climatic factors on them. Sci Hortic 250:344–351
20. Chapagain AK, Hoeksta AY, Savenije HHG (2006) Water saving through international trade of agricultural products. Hydrol Earth Syst Sci 10:455–468. https://doi.org/10.5194/hess-10-455. https://hal.archives-ouvertes.fr/hal-00298749/
21. Ababaei B, Ramezani Etedali H (2017) Water footprint assessment of main cereals in Iran. Agric Water Manag 179:401–411. http://www.sciencedirect.com/science/article/pii/S03783774163 02621
22. Karshenas M (1990) Oil, state and industrialization in Iran. CUP Archive
23. Ministry of Agriculture- Jihad (MAJ) (2017). http://www.maj.ir/Portal/Home/Default.aspx?CategoryID=c5c8bb7b-ad9f-43dd-8502-cbb9e37fa2ce
24. Bazrafshan O, Zamani H, Etedali HR, Moshizi ZG, Shamili M, Ismaelpour Y, Gholami H (2020) Improving water management in date palms using economic value of water footprint and virtual water trade concepts in Iran. Agric Water Manag 229:105941
25. Fulton J, Norton M, Shilling F (2019) Water-indexed benefits and impacts of California almonds. Ecol Ind 96:711–717
26. Vafaei K, Bazrafshan O, Ramezani Etedali H (2020) Spatial and temporal changes of ecological water footprint and virtual water trade in irrigated and rain-fed almond production at Iran. JWSS 24(2):287–302 (In Persion)
27. Ministry of Agriculture-Jihad (MAJ) (2018). http://www.maj.ir/Portal/Home/Default.aspx?CategoryID=c5c8bb7b-ad9f-43dd-8502-cbb9e37fa2ce

Appraising the Water Status in Egypt Through the Application of the Virtual Water Principle in the Agricultural Sector

Mohannad Alobid and István Szűcs

Abstract Egypt has a serious problem with its water resources, which is caused by a mismatch between rising water demand and available supply. As a result, Egypt must examine a number of elements and/or factors, including the water footprint and the virtual water, in order to preserve water and attain food security. In this research it was found that Egypt is an importer of virtual water and not an exporter of maize and wheat crops, and for the rice crop Egypt exports an important amount of their productions as evidenced by the amount of exported virtual water. Furthermore, the agricultural sector consumes approximately 33 billion cubic meters of "blue water" and 6.5 billion cubic meters of "green water," which is less than the number of renewable water resources 58.5 billion cubic meters per year. The reason for this is that the volume of agricultural sector consumption was calculated using the volume of virtual water. Additionally, it was noticed that the amount of water used in irrigation operations (blue water) was calculated from the available water resources in irrigation operations, which is about 34 billion cubic meters, and the rest is about 11 billion cubic meters of rainwater (green water). Moreover, Egypt's reliance on external resources to meet its agricultural crop demands is estimated about 21.15%, which is not a huge number, and its reliance on its water resources and Water Import Dependency Index (WIDI) is estimated to be around 78.84%. Also, Egypt has a high rate of food Self-Sufficiency Ratio in terms of (rice, potatoes, vegetables, and fruits), with scores of (94.9, 94.84, 99.3, 99.92, and 101.1), respectively, resulting in less reliance on external water resources to meet food needs for its people.

Keywords Water footprint · Virtual water · Food security · Blue water and WIDI

M. Alobid (✉) · I. Szűcs
Faculty of Economics and Business, Institute of Applied Economic Sciences, University of Debrecen, 4032 Debrecen, Hungary
e-mail: mohannad.alobid@econ.unideb.hu

© The Author(s), under exclusive license to Springer Nature Singapore Pte Ltd. 2022
S. S. Muthu (ed.), *Environmental Footprints of Crops*,
Environmental Footprints and Eco-design of Products and Processes,
https://doi.org/10.1007/978-981-19-0534-6_2

1 Introduction

In the last thirty years, many studies have focused on the need to protect available environmental resources and combat climate change [21]. As a result, new ideas for conserving water rather than wasting it have been devised. The notions of virtual water and water footprint are among these concepts, which have been used not only in agriculture and industry but also in the health sphere [12, 15], such as changing food patterns as a required precursor for health prevention and health improvement [26].

Furthermore, in order to confront the world's impending threat as a result of climate change, researchers are currently developing new ideas by turning to renewable energies such as solar energy and gradually dispensing with traditional energy sources [16], as well as studying the challenges of transitioning from a fossil-fuel-based economy to a bio-economy and the effects on food production, feed, bioenergy, and other vital materials [22]. In this study, the concepts of virtual water and water footprint indicators were shed light on as a case study of Egypt.

Egypt deteriorates from a critical shortage of water resources that figures in the imbalance between the increase of water demand and the availability of the accessible water. To solve this problem, it was necessary to coordinate with the ten Nile Basin countries, to ensure an abundant water future [19].

Egypt depends on 97% of its water needs on the Nile River. The minimum rainfall is 18 mm per year, and most of it falls during the fall and winter seasons. In 1959, the Nile Water Treaty was concluded between Egypt and Sudan, allocating 55.5 billion cubic meters of water annually to Egypt, without specifying any allocation to the upstream countries that are located next to Sudan 18.5 billion cubic meters per year [28]. There was no agreement to share water between all the ten riparian states of the Nile. However, the Riparian countries are cooperating through the Nile Basin Initiative [13]. The management of water resources in Egypt depends on a complex set of infrastructure along the river. Egypt's water resource management is reliant on a complex network of infrastructure along the Nile. The Aswan High Dam, which forms Lake Nasser, is the most essential part of this infrastructure. Egypt is protected from floods by the High Dam, which also stores water for year-round irrigation and generates hydropower. After the downstream of the river from the Aswan Dam, there are seven canals to increase the river's water level, so that it can flow into irrigation channels from the first level as shown in Fig. 1.

In addition, Fig. 2 shows the water balance between water supply and water loss as an average water capacity of the Nile river (billion m^3/year) [2, 11].

Providing water to the agricultural sector is one of the main strategic objectives that Egypt aims to secure enough water to serve the population whose numbers are increasing as resources remain limited. But the volume of the potential savings in water and agriculture, and how best to achieve a breakthrough on such a topic, have caused some controversy [17]. While the irrigation efficiency at the field level may be low due to the predominance of flood irrigation, the public system efficiency is generally high due to the return of flows. Water-saving irrigation equipment such

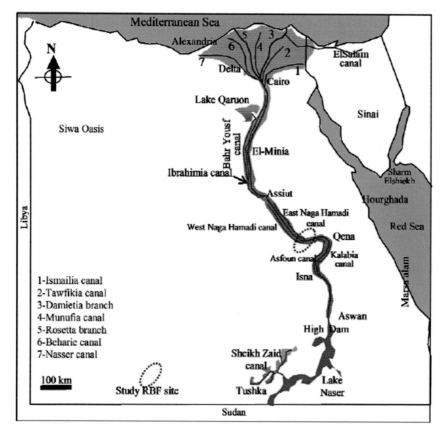

Fig. 1 Map of Egypt with Nile River. *Source* [23]

as sprinklers and drip irrigation are not a top priority in Egypt's water conservation measures. It is instead reliant on farmer expertise, which lacks the ability to forecast when and how much water will be available. They irrigate near each other and waste a lot of water. As a result, after understanding the problem in relation to water and population expansion, Egypt must consider a number of aspects, including the water footprint and virtual water, in order to track the water's fate and attain food security. Among other things, we began by calculating virtual water for the most important agricultural crops and products, then assessing Egypt's water footprint and indicators, evaluating food security, and estimating food self-sufficiency for a few selected crops and products, as described in the following chapters.

The main objective and research question (RQ)

The main objective of the study is to benefit from the principle of virtual water for:

- knowing the true magnitude of the water deficit in Egypt;

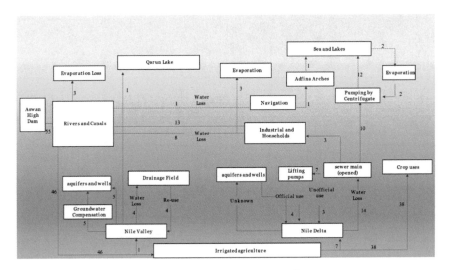

Fig. 2 The average water capacity of the Nile River (billion m³/year). *Source* Own constructed from [2, 11]

- using water more efficiently in the agricultural sector. To achieve the goals of this research, the following has applied:
- Calculating the virtual water for the most important agricultural crops in Egypt;
- Calculating the virtual water for the agricultural products;
- Assessment of the water footprint and its indicators in Egypt;
- Evaluating the food security and estimating food self-sufficiency for some selected crops and products.

RQ: *How to use the principles of virtual water and water footprint indicators to estimate the volume of water deficit in Egypt and to achieve water efficiency consumption in the agricultural sector?*

2   Material and Methods

The main sources of the data to assess the water footprint, virtual water for some selected crops and products in Egypt and the value and quantity of the food gap in Egypt during 2000–2018 are mainly based on the National Water Footprint Website, the Central Agency for Public Mobilization and Statistics (CAPMAS) in Egypt, the Ministry of Water Resources and Irrigation in Egypt, and the Ministry of Agriculture in Egypt. In addition, the average world market prices USD/ton for the crops was collected from the World Bank Open Data Resource and "Water footprints of nations Volume 2: Appendix" [3].

2.1　Calculating the Virtual Water for Agricultural Crops

The amount of virtual water had calculated in this research for most important crops grown in Egypt and for the main plant products (oils and refined sugar) Eq. 1. As for calculating the amount of virtual water in agricultural crops, it was done according to the following relationship:

$$VWC(c) = \frac{CWU(c)}{production(c)} \tag{1}$$

where

VWC: is the virtual water C of a crop (m^3/ton),
CWU: amount of water consumed by the crop C m^3/year,
Production: production in tons of yield C,

The crop water use for each crop included in the study was calculated with the following Eq. 2:

$$CWU(c) = CWR(c) \times \frac{production(c)}{Yield(c)} \tag{2}$$

where

CWU: amount of water consumed by the crop C m^3/year.
CWR: the amount of water requirements for each crop (c) measured in the field in m^3/ ha. It is defined as the amount of water required for evapotranspiration from the planting until the harvest for a specific crop that grows in soil containing sufficient water for it.
Production: production in tons of yield C.
Yield: yield of crop (c) per unit area, measured in tons/ha.
The quantity of water requirements of crop (c) is calculated from the following relationship Eq. 3:

$$CWR = 10 \times \sum_{d=1}^{lp} ET_c(c, d) \tag{3}$$

where

CWR: the amount of water requirements for each crop (c) measured in the field in m^3/ha. It is defined as the amount of water required for evapotranspiration from the planting until the harvest for a specific crop that grows in soil containing sufficient water for it.

The factor 10 is meant to convert mm into m^3/ha, and the summation is done over the period from the first to the final day of the growing period.

lp: represents the length of growth, measured in days.

Fig. 3 Crop growth stages
for different types of crops.
Source [1]

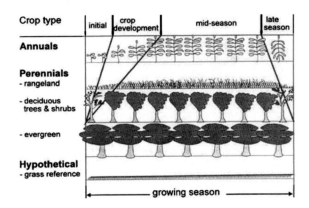

Etc: is the amount of daily evapotranspiration of the crop (c) and it is measured in mm. This evapotranspiration is obtained by the process of multiplying the reference evapotranspiration amount Eto by the coefficient of the crop Kc. The crop coefficient is taken from four stages of the crop growth; initial, crop development, mid-season, and late season Fig. 3, that is the stage where the crop is ready for harvest Equation (4) [1, 5].

$$\mathbf{ETc_c = Kc_c \times ET_0} \tag{4}$$

$\mathbf{ET_0}$ the amount of reference evapotranspiration, which is the percentage of evapotranspiration from the grass in specific growth conditions, is affected by climatic conditions only [7].

The water consumption of crops varies according to their growth stage. In fact, the water consumption is at a low rate at the beginning of the crop growing season, when it is mostly in form of evaporation from the soil surface, then it increases with the plant growth as a result of the leaf mass surface increase and becomes in form of transpiration from the leaves up to the maximum growth stages [1]. The crop coefficient (Kc) indicates the relationship between the evapotranspiration of the crop (ETc) and the reference evapotranspiration (ET$_0$). Kc differs according to the type of the crop, the growth phase, the growing season, and the associated climatic conditions. Kc expresses the effect of the properties that distinguish the field crop from the reference grass, whose appearance is stable and covers the entire ground, and therefore different crops have different Kc [25].

Several weather conditions effects have been incorporated into the ET$_0$, which then represents an indicator of the atmospheric requirements needed for the process of evapotranspiration from green grass surfaces. Accordingly, the crop coefficient (Kc) varies greatly with the characteristics of the crop and to a small extent with the climate. The thing that explains the possible transfer of the crop coefficient (Kc) values that were calculated at one of the irrigation research stations for generalization between sites and climatic regions by [6] to the (ET$_0$) reference that represents the incorporation of the effects of four basic characteristics distinguishing the evapo-transpiration of the crop from the reference evapotranspiration according to [1, 27], namely,

- Resistance to air movement and the turbulent transfer of vapor from the crop to the atmosphere, in addition to leaf properties, stomata, and spacing between plants.
- Reflected radiation albedo, which is affected by the part of the earth covered by vegetation, and the amount of wetness of the soil surface, which is the main source of energy exchange for the evaporation process.
- Crop resistance to heat transfer and its relationship to leaf area, number of pores, age, and condition.
- The amount of wetness of the soil surface and the portion of the ground covered by vegetation on the surface resistance.

The ET_0 was calculated as an average of all the months of the year through the CROPWAT program. As for the Kc values, it was derived from a previous study performed by the researchers [1, 3].

2.2 Calculation of the Virtual Water for Agricultural Products

The virtual water for plant products in m^3/ton Eq. 5:

$$VWCp(p) = VWC(c) \times \frac{vf[p]}{pf[p]} \tag{5}$$

where
pf is the production factor that indicates the weight of the primary product resulting from one ton of the main crop;

vf is the value factor (USD/ton), and the product value in USD is the sum of the product values resulting from this main crop.

As for the secondary products from crops, they are calculated from the following relationship Eq. 6:

$$VWCp(p) = VWC(primary\ product) \times \frac{vf[p.p]}{pf[p.p]} \tag{6}$$

where
pf is the production coefficient that indicates the weight of the secondary agricultural product resulting from one ton of the primary product;

Vf is the value coefficient (USD/ton), and it is the sum of the value of the secondary product to the sum of the values of the products produced from the primary product [10].

2.3   Water Footprint and Its Indicators

The concept of a country's water footprint is defined as the total freshwater volume used in service sectors and in the production of consumed products of all kinds by that country.

This concept was discovered by the researcher HOEKESTRA in 2002 [9] in order to determine the actual water consumption per capita or country and give real information for water consumption other than the traditional information about the quantities of surface and groundwater withdrawal used in the agricultural, industrial, and domestic sectors usually in calculating the annual water balance and from them, the quantities of water actually used are greater than the quantities of withdrawal from local ground and surface water, hence the concept of virtual water imported or exported, and then the so-called virtual water trade between countries.

The water footprint consists of two parts, as in the following relationship Eq. 7:

$$WFP = IWFP + EWFP \tag{7}$$

where
IWFP: Internal Water Footprint;
EWFP: External Water Footprint.
As for the **IWFP**, it is calculated from the following relationship Eq. 8:

$$IWFP = DWW + IWW + AWU - VWE \tag{8}$$

As **DWW** is the amount of water withdrawals for the domestic sector, **IWW** is the amount of water withdrawals for the industrial sector depending on the principle of virtual water and **AWW** is the amount of water consumption in the agricultural sector and is calculated depending on the method of calculating the amount of virtual water for crops and agricultural products as previously explained and **VWE** is the amount Water exported through agricultural products to other countries.

External water footprint calculated from the following relationship Eq. 9:

$$EWFP = VWI - VWE\ re\_export \tag{9}$$

where
VWI is the volume of the imported virtual water;
VWE re-export: the volume of virtual water re-exported from imported products.
In this research, the water footprint was calculated based on the calculation of the water included in agricultural products only and did not include animal and industrial products to give an initial picture of the true water scale if the water included in the products was taken into account when calculating the annual water balance.

2.3.1 Water Footprint Indicators

(1) **Water Import Dependency Index (WIDI):** which is equal to the ratio of the external water footprint on the total water footprint as shown in the following relationship Eq. 10:

$$WIDI = \frac{EWFP}{WFP} x100 \qquad (10)$$

(2) **Water Self-sufficiency Index (WSSI):** which is equal to the ratio of the internal water footprint to the total water footprint. It is calculated as relationship Eq. 11:

$$WSSI = \frac{IWFP}{WFP} x100 \qquad (11)$$

This indicator is 100% if the available water within the country meets all needs in all consumption and product sectors [3].

2.4 Food Security and Food Self-sufficiency in Egypt

Most countries are trying hard to achieve food security locally without relying on external resources and food imports. It must be pointed out the close relationship between food security and water security, as the volume and quality of available water will have a negative impact on food production and hence on food security. Water security is defined as the ability to meet all water needs in all sectors of water use with the necessary quantity and quality [14].

As for food security, it is defined as the ability of production to adequately fill food while increasing stability in production processes and ensuring food access to all citizens naturally and economically [4, 8]. From previous definitions we can define the food gap which is the difference between available supply and consumption in the country. The data were estimated and the food gap was conducted from different resources mentioned later. The **Self-sufficiency Ratio (SSR)** is defined as Eq. 12:

$$\textbf{SSR} = \textbf{Production} \times \textbf{100}/(\textbf{Production} + \textbf{Imports} - \textbf{Exports}). \qquad (12)$$

In this research, only agricultural crops, sugar, and oils were discussed according to the calculations of the volume of virtual water.

Table 1 The virtual water volume for most important crops

Crop	Volume of virtual water m³/ton
Wheat	728.56
Rice	1025.14
Maize	1072.31
Potato	330.59

Source Own calculation (2021)

3 Results and Discussions

3.1 Virtual Water for Agricultural Crops

The amount of crop evapotranspiration Etc (mm), crop water requirements CWR (m³/hectare) production (tons/year), productivity (kg/hectare), water use by crop $CWU_{[c]}$ (m³/year), virtual water requirements for crops $Vwc_{[c]}$ (m³/ton) have calculated for the studied crops, and Appendix 1 shows the volume of those variables, and Table 1 shows the virtual water volume m³/ton for most important crops.

3.2 The Virtual Water for Agricultural Products

Table 2 shows the volume of virtual water for plant oils (olive, soybean, and cotton seeds) as a primary product and for sugar as a by-product.

Table 2 The virtual water volume for some selected plant products

Plant oils produced by	Volume of virtual water m³/ton
Olive	37,274
Soybean	12,490
Cotton seeds	4091
Average	17,952
Refined sugar	1426

Source Own calculation (2021)

3.3 Indicators of Water Footprint

3.3.1 Volume of Exported and Imported Water and Water Balance in Egypt

Table 3 shows the volume of water included in the exported and imported agricultural products only for three strategic crops grown for the years 2000–2018.

It is clear that Egypt is an importer of virtual water and not an exporter of maize and wheat crops, and for the rice crop Egypt exports an important amount of their productions which is appear from the amount of exported virtual water. Based on the calculation of the volume of virtual water included in crops and plant products, the water balance of Egypt was calculated during the years 2000–2018 as shown in the following Table 4.

From the results of the previous table, it was found that the volume of real water consumption in Egypt is about 24.53 billion cubic meters/year, only for the three major crops and the volume of real individual consumption in the years between

Table 3 The volume of virtual exported and imported water for the three main crops (rice, maize, and wheat)

The volume of virtual water	Egypt
Exported million m³/year	1040
Imported million m³/year	17,300
The difference between imported and exported Billion m³/year	−16,260

Source Own calculation (2021)

Table 4 The water balance in Egypt using the concept of virtual water

The water balance in Egypt using the concept of virtual water		
Renewable water billion m³/year		58.5
Population/million		98.42
Water uses million m³	Households	9000
	Agriculture	67,000
	Industry	2000
Exported VW in agricultural sector million m³ for (rice, maize, and wheat)		1040
Imported VW in agricultural sector million m³ (rice, maize, and wheat)		17,300
Internal water footprint IWFP billion m³ (rice, maize, and wheat)		19.34
External water footprint EWFP billion m³ for (rice, maize, and wheat)		5.19
The Total water footprint WFP billion m³ for (rice, maize, and wheat)		24.53

Source Own calculation (2021)

2000–2018 is about 232.49 cubic meters/year for those crops. It is clear that the consumption of the agricultural sector is about 33 billion cubic meters "blue water" and around 6.5 billion cubic meters "green water", which are less than the number of renewable water resources 58.5 billion cubic meters per year. The explanation for this is that the calculation of the volume of consumption of the agricultural sector was based on the volume of virtual water for agricultural products. It is known that a high percentage of crops in Egypt depend on irrigated water, the total cultivated area is 7.2 million feddans (1 feddan = 0.42 ha), representing only 3% of the total land area. The entire crop area is irrigated, except for some rain-fed areas on the Mediterranean coast according to the estimates of the Arab Organization for Agricultural Development and Irrigated Agriculture in Egypt [18, 24], meaning that Egypt is one of the countries that relies heavily on the waters of the Nile River in agriculture.

Depending on the percentage of irrigated land from the total agricultural lands, the amount of water used in irrigation operations (blue water) is calculated from the available water resources in irrigation operations, which is about 34 billion cubic meters, and the rest is about 11 billion cubic meters of rainwater (green water) as shown in Fig. 4.

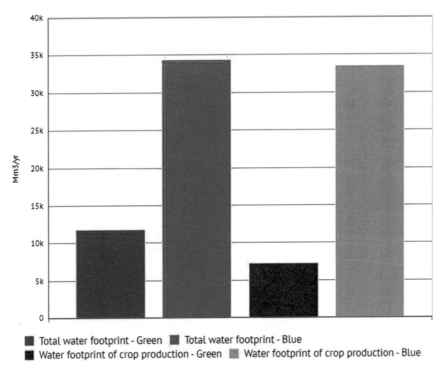

Fig. 4 The consumptions of green and blue water in the agricultural sector. *Source* The National Water Footprint Statistics (https://knoema.com/WFPNWFPS2015/national-water-footprint-statis tics?location=1000590-egypt&action=export&gadget=visualization)

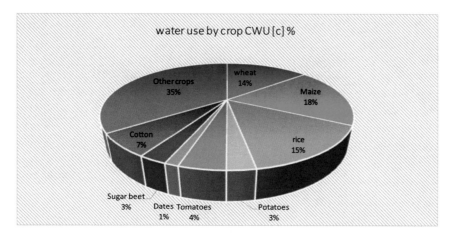

Fig. 5 The percentages of water distribution consumed in agriculture among crops. *Source* Own calculation (2021)

From Fig. 5 it turns out that 18% of the water consumed in agriculture (which is about 45 billion cubic meters according to the foregoing) is consumed from maize, which is one of the most important strategic crops that the Egyptian government supports its cultivation and it is one of the crops that depend on irrigation mainly and then rice 15% wheat and 14%, then cotton 7% (especially the last 10 years because the government support farmers in planting wheat and maize other than cotton) and other crops 35%.

The increase in agricultural production Egypt as most Arab countries "especially after the Arab spring" does not meet the increasing demand for foodstuffs due to the high population growth rate and the most important factors impeding the growth of agricultural production in Egypt are mainly water, desertification, and the lack of arable land [29].

3.3.2 Water Import Dependency Index and Water Self-sufficiency Index

Table 5 shows the value of these two indicators in Egypt, based on the previous calculations of the volume of virtual water in main crops and in the exported and imported agricultural products.

From the previous Table 5, it turns out that the extent of Egypt dependence on external resources to meet its agricultural crops needs is about 21.15%, and this is not a large percentage, and its dependence on its water resources and self-sufficiency is about 78.84%.

Table 5 The water footprint indicators in Egypt

Denomination	Value
The renewable water in Egypt billion m^3/year	58.50
Exported water in agricultural sector million m^3 for (rice, maize, and wheat)	17,300.00
Imported water in agricultural sector million m^3 for (rice, maize, and wheat)	1040.00
Internal Water Footprint (IWFP) billion m^3 (rice, maize, and wheat)	19.34
External Water Footprint (EWFP) billion m^3 for (rice, maize, and wheat)	5.19
The Total Water Footprint (WFP) billion m^3 for (rice, maize, and wheat)	24.53
Water Import Dependency Index (WIDI) %	21.15
Water Self-sufficiency Index (WSSI) %	78.84

Source Own calculation (2021)

3.4 Food Security and Food Self-sufficiency

As mentioned in the material and method the food gap is the difference between available supply and consumption. Appendix 2 shows food gap of Egypt during the period 2000–2018 as an example for some selected crops (wheat, rice, and maize). Egypt suffers from food gap in various agricultural food commodities (except rice).

The lack of development of agricultural production in most Arab countries results in dependence on external resources to bridge the deficit in food products, especially the main ones.

Relying on the data mentioned on Table 6 in estimating the average value of the food gap in relation to the main crops in the food balance in Egypt for the years 2000–2018 as shown in Table 6.

The value of the food gap is affected by the fluctuation of international food prices and food support policies, as well as the change in food reserves and the stock of exporting countries.

This research focused on the main elements of the food balance in relation to crops to show the food and water demands to meet this need and the extent to which food

Table 6 The value and quantity of the food gap in Egypt during 2000–2018

Food gap	Egypt
The Quantity million tons	1934
The value million USD	17,987
Cereals million tons	1221
Cereals million USD	3,539
Population (1000)	98,420
The value of food gap per capita (USD/year)	560

Source Collected and calculated from different resources (FAO, FAOSTAT online database), (http://agri.sprograming.com/), (https://www.indexmundi.com/) and (http://www.aoad.org/)

Table 7 Self-sufficiency ratio through the main elements in the food balance in Egypt

Products	Self-sufficiency ratio %
Wheat	52.81
Maize	59.47
Rice	94.9
Barley	65.34
Potatoes	99.3
Legumes	7.6
Vegetables	99.92
Fruits	101.1
Refined sugar	69.6
Plant Oils	11.3

Source Calculated from (http://agri.sprograming.com/) and (http://www.capmas.gov.eg/)

security can be achieved from local production and by relying on available water resources.

The main elements of the food balance, according to the estimates of the World Food Organization, are wheat, maize, rice, legumes, vegetables, fruits, sugar, plants oil, meat (red meat, white meat), white fish, milk, and dairy products.

As mentioned previously, Table 7 shows the self-sufficiency ratio of the main crops in the food balance of the Egyptian Arab Republic for the years 2000–2018 see also Appendix 3.

Egypt has a high rate of self-sufficiency in relation to (rice, potatoes, vegetables, and fruits) which are (94.9, 94.84, 99.3, 99.92, and 101.1) respectively, and this has led to a decrease in the degree of dependence on external water resources to meet food needs. On the other hand, we can see that Egypt has a low rate of self-sufficiency in (wheat, maize, legumes, barley, and plants oil) with ratio (52.81, 59.47, 7.6, 65.34, and 11.3), respectively, and this has led to an increase in the degree of dependence on the external water resources to meet food demand in the country.

3.5   The Volume of Virtual Water Required for Self-sufficiency

Water, and especially irrigations, has an important effect on food production, especially in countries that depend on irrigations for agriculture, such as Egypt, where the percentage of irrigated lands exceeds 95% of the total cultivated land [20].

The amount of food requirement of each element in the food balance was calculated during the years 2000–2018 based on the data of the World Food Organization and only by knowing the quantity of import and export from each component as well

Table 8 The volume of virtual water required to stopping food gap for the main crops in Egypt

Products	The quantity needed 1000 ton	Required water needed million m^3
Wheat	7193.27	728.56
Maize	4878.47	1072.31
Rice	374.62	1023.16
Barley	6.84	1940.35
Potatoes	−290.74	−329.30
Legumes	17.36	37.25
Vegetables	−6.89	−18.67
Fruits	−30.62	−79.74
Refined sugar	3650	6741
Plant Oils	212	3266

Source Collected and calculated depends on the data from FAO database (http://www.fao.org/fao stat/en/#data/QC)

as local production see Appendix 4 (A, B), and then the volume of virtual water for the crops mentioned in the previous table was calculated.

The main crops imported in Egypt to bridge the food gap are wheat, maize, and barley, and the most important exports are cotton, rice, and tomato.

Table 8 shows the volume of water needed to meet the food requirement of the main elements in the food balance and to obtain self-sufficiency from these elements locally.

The negative sign in Table 8 indicates that it had used an additional amount of water to produce a higher amount of the crop than the appropriate food requirement.

4 Conclusion

The virtual water principle is a good tool in managing water resources due to its close connection with the water footprint principle because it helps in determining the true water balance for Egypt, and it helps in trying to provide water for more economic uses than agriculture. It was found that Egypt is an importer of virtual water and not an exporter of maize and wheat crops, and for the rice crop Egypt exports an important amount of their productions which is appear from the amount of exported virtual water. In addition, the consumption of the agricultural sector is about 33 billion cubic meters of "blue water" and around 6.5 billion cubic meters of "green water," which are less than the number of renewable water resources 58.5 billion cubic meters per year. The explanation for this is that the calculation of the volume of consumption of the agricultural sector was based on the volume of virtual water for agricultural products. Furthermore, it was noticed that the amount of water used in irrigation operations (blue water) was calculated from the available water resources in irrigation operations, which is about 34 billion cubic meters, and the rest is about 11 billion cubic meters of rainwater (green water). Additionally, as it turns out that the extent of Egypt dependence on external resources to meet its agricultural crops needs is about 21.15%, and this is not a large percentage, and its dependence on its water resources and self-sufficiency is about 78.84%. Also, Egypt has a high rate of self-sufficiency in relation to (rice, potatoes, vegetables, and fruits) which are (94.9, 94.84, 99.3, 99.92, and 101.1), respectively, and this has led to a decrease in the degree of dependence on external water resources to meet food needs for the country.

Appendices

Appendix 1: The Virtual Water Volume for Selected Crops

Crop	Crop Evapo-transpiration Etc (mm)	Crop Water Requirements CWR m^3/hectare	Production tons/year	Productivity kg/ha	Water use by crop CWU [c] m^3/year	Virtual water requirements for crops Vwc [c] m^3/ton
Wheat	492	4912	8,052,105.26	6742.08	5,866,430.10	728.56
Barley	456	4562	125,710.47	2351.12	243,922.54	1940.35
Maize	831	8312	7,160,365.53	7751.47	7,678,151.15	1072.31
Rice	1034	10,346	6,045,263.16	10,111.81	6,185,271.74	1023.16
Potatoes	848	8487	3,480,719.42	25,772.47	1,146,217.87	329.30
Sugar cane	1687	17,120	15,929,216.84	116,848.15	2,333,868.29	146.51
Sugar beet	846	8460	6,765,547.11	50,180.84	1,140,605.23	168.59
Broad beans, horse beans, dry	481	4814	243,489.79	3332.21	351,766.50	1444.69
String beans	562	5628	278.16	6721.44	232.91	837.32
Chickpeas	428	4281	7618.74	2048.57	15,921.27	2089.75
Lentils	881	8814	1917.74	1899.64	8897.98	4639.83
Sesame seed	595	5950	39,848.53	1290.87	183,673.61	4609.29
Olive	1083	10,837	507,148.63	8686.01	632,738.13	1247.64
Soybeans	978	9782	32,346.53	3173.29	99,711.58	3082.61
Tomatoes	916	9164	7,854,596.21	38,613.73	1,864,091.34	237.32
Onions, dry	1007	10,073	1,751,625.16	32,503.97	542,829.70	309.90
Watermelons	648	6489	1,668,401.16	28,803.03	375,872.09	225.29
Melons, other (inc. cantaloupes)	601	6012	826,307.95	24,461.97	203,081.08	245.77
Peas, green	646	6465	234,780.79	9976.85	152,137.98	648.00
Peas, dry	714	7148	199.16	1873.69	759.78	3814.93

(continued)

(continued)

Crop	Crop Evapo-transpiration Etc (mm)	Crop Water Requirements CWR m³/hectare	Production tons/year	Productivity kg/ha	Water use by crop CWU [c] m³/year	Virtual water requirements for crops Vwc [c] m³/ton
Cabbages and other brassicas	925	9250	592,925.32	30,100.09	182,210.72	307.31
Cucumbers and gherkins	783	7833	574,737.11	21,445.53	209,923.27	365.25
Pumpkins, squash, and gourds	602	6022	610,512.74	18,168.72	202,353.70	331.45
Artichoke	1486	14,860	188,478.58	20,729.18	135,113.48	716.86
Beans, green	560	5608	262,763.68	10,475.51	140,668.92	535.34
Carrots and turnips	461	4617	177,560.79	28,911.67	28,355.27	159.69
Garlic	958	9586	244,759.89	23,515.83	99,774.00	407.64
Beans, dry	662	6627	77,613.21	2766.89	185,892.01	2395.11
Okra	1013	10,133	88,126.79	13,733.99	65,020.34	737.80
Lettuce and chicory	617	6172	115,418.95	23,386.88	30,460.06	263.91
Berries	1540	15,405	596.32	7292.98	1259.61	2112.31
Dates	1458	14,581	1,323,964.53	34,264.75	563,399.03	425.54
Figs	1040	10,406	199,124.84	6357.76	325,915.59	1636.74
Pears	1348	13,480	51,790.53	14,948.8	46,701.83	901.74
Apple	1348	13,480	575,130.21	22,755	340,705.57	592.40
Grapes	831	8312	1,414,360.63	21,380.33	549,858.94	388.77
Peaches and nectarines	1348	13,480	326,216.26	11,253.95	390,742.38	1197.80
Oranges	1040	10,407	2,410,495.05	23,098.44	1,086,048.32	450.55
Lemons and limes	1040	10,407	322,586.89	20,318.62	165,225.87	512.19
Bananas	1747	17,471	188,478.58	43,722.86	75,313.22	399.59
Lupins	607	6075	2435.11	1861.89	7945.31	3262.81
Groundnuts, with shell	1550	15,509	203,457.58	3268.12	965,516.45	4745.54
Cauliflowers and broccoli	857	8571	129,018	26,833.25	41,210.56	319.42
Linseed	816	8160	14,642.05	1655.82	72,157.08	4928.07
Cotton	966	9667	491,432.58	2732.89	1,738,335.48	3537.28
Nuts	1550	15,503	12,383.63	5089.27	37,723.17	3046.21
Anise, Fennel, and other aromatic plants	841	8412	24,565.47	865.25	238,826.62	9722.05
Chillies and peppers, dry	855	8550	52,159.53	3172.46	140,573.56	2695.07

(continued)

(continued)

Crop	Crop Evapo-transpiration Etc (mm)	Crop Water Requirements CWR m³/hectare	Production tons/year	Productivity kg/ha	Water use by crop CWU [c] m³/year	Virtual water requirements for crops Vwc [c] m³/ton
Chillies and peppers, green	842	8427	588,170.32	16,248.67	305,041.05	518.63
Eggplants (aubergines)	832	8323	1,143,314.74	25,507.25	373,062.90	326.30
Fruit, tropical fresh	1246	12,467	19,356.32	7845.76	30,757.41	1589.01
Fruit, citrus	1040	10,406	4304.05	14,675.89	3051.80	709.05
Fruit, fresh	1245	12,458	419,430.11	18,483.31	282,701.55	674.01
Mangoes, mangosteens, guavas	1347	13,471	679,717.74	9925.06	922,561.44	1357.27
Sorghum	760	7603	827,291.32	5451.87	1,153,713.48	1394.57
Spinach	926	9267	42,717	16,370.4	24,181.35	566.08
Strawberries	1341	13,476	210,111.05	33,855.28	83,634.12	398.05
Sweet potatoes	846	8460	333,758.53	28,999.85	97,365.92	291.73

Appendix 2: Consumption and Food Gap for Wheat, Rice and Maize Crops

Appendix 3A: Self-sufficiency Ratio SSR for the Wheat Crop

	Year/wheat	Productions tons	Export tons	Imports tons
SSR =	2000	6,650,000	880	4,896,000
	2001	6,420,000	945	4,413,000
	2002	6,790,000	784	5,575,000
	2003	6,820,000	1120	4,057,000
	2004	7,160,000	864	4,363,000
	2005	8,150,000	457	5,688,000
	2006	8,270,000	264	5,817,000
	2007	7,390,000	7150	5,911,000
	2008	7,970,000	3520	5,205,000
	2009	8,530,000	4580	4,092,000
	2010	7,170,000	4530	9,804,780
	2011	8,380,000	2786	9,800,060
	2012	8,790,000	265	6,537,632
	2013	9,460,000	34	7,869,653
	2014	9,270,000	599	8,515,058
	2015	8,670,000	498	8,981,777
	2016	9,570,000	1390	10,788,295
	2017	8,410,000	1125	12,025,245
	2018	9,120,000	4766	12,369,230
	average	8,052,105	1924	7,195,196
	52.8167			

Appendix 3B: Self-sufficiency Ratio SSR for the Maize Crop

	Year/Maize	Productions tons	Export tons	Imports tons
SSR =	2000	6,474,450	2110	3,968,020
	2001	6,093,578	1247	4,945,481
	2002	6,430,962	941	3,876,921
	2003	6,530,427	2547	4,128,741
	2004	6,236,140	965	3,978,450
	2005	7,085,190	1478	4,268,746
	2006	6,374,300	490	3,958,740

(continued)

(continued)

Year/Maize	Productions tons	Export tons	Imports tons
2007	6,243,220	1550	4,428,310
2008	7,401,412	1010	2,463,190
2009	7,686,091	930	1,872,520
2010	7,041,099	16,080	4,844,481
2011	6,876,473	753	6,861,685
2012	8,093,646	2100	3,131,351
2013	7,956,593	306	5,738,431
2014	8,059,906	405	4,326,802
2015	7,803,183	270	6,779,475
2016	7,817,640	482	6,036,522
2017	8,542,635	340	8,703,411
2018	7,300,000	702	8,414,392
average	7,160,366	1827	4,880,298
59.477218			

Appendix 4: Required Water Needed for Crops (million m³)

Crop	Crop Evapo-transpiration Etc (mm)	Crop Water requirements CWR m³/hectare	Production tons	Productivity kg/hectare	Water use by crop CWU [c] m³/year	Virtual water requirements for crops Vwc [c] m³/ton
Wheat	492	4912	7,193,270	6742.08	5,240,718.33	728.56
Barley	456	4562	6840	2351.12	13,272.00653	1940.35
Maize	831	8312	4,878,470	7751.47	5,231,245.511	1072.31
Rice	1034	10,346	-374,620	10,111.81	-383,296.217	1023.16
Potatoes	848	8487	-290,740	25,772.47	-95,742.09922	329.30
Fruits	3531	35,331	-30,620	443,090.48	-2441.567284	79.74

References

1. Allen RG, Pereira LS, Raes D, Smith M (1998) Crop evapotranspiration-guidelines for computing crop water requirements-FAO irrigation and drainage paper 56. FAO Rome 300(9):D05109
2. Ashour M, El Attar S, Rafaat Y, Mohamed M (2009) Water resources management in Egypt. JES J Eng Sci 37(2):269–279
3. Chapagain AK, Hoekstra AY (2004) Volume 2: appendices (No. 16). Research report series 'value of water. Chapagain AK, Hoekstra AY (2004) Water footprints of nations
4. Rome declaration on world food security and world food summit plan of action: world food summit, 13–17 Nov 1996, Rome, Italy
5. Droogers P, Allen RG (2002) Estimating reference evapotranspiration under inaccurate data conditions. Irrig Drain Syst 16(1):33–45
6. Hargreaves GH, Merkley GP (1998) Irrigation Fundamentals: an applied technology text for teaching irrigation at the intermediate level. Water Resources Publication
7. Hargreaves GH, Samani ZA (1985) Reference crop evapotranspiration from temperature. Appl Eng Agric 1(2):96–99
8. Hendriks SL (2015) The food security continuum: a novel tool for understanding food insecurity as a range of experiences. Food Secur 7(3):609–619
9. Hoekstra AY (2003) Virtual water trade: a quantification of virtual water flows between nations in relation to international crop trade. In: Proceedings of the international expert meeting on virtual water trade 12, Delft
10. Hofwegan PV (2003) Virtual water- conscious choices. World Water Council
11. Land F (1997) Irrigation potential in Africa; a basin approach. 4
12. Mbow C, Rosenzweig C, Barioni LG, Benton TG, Herrero M, Krishnapillai M, Liwenga E, Pradhan P, Rivera-Ferre M-G, Sapkota T (2019) Food security. In: Climate change and land, pp 437–550
13. Mekonnen DZ (2010) The Nile Basin cooperative framework agreement negotiations and the adoption of a 'water security'paradigm: flight into obscurity or a logical cul-de-sac? Eur J Int Law 21(2):421–440
14. Mouailli F (2005) The water security problem. Researches and reports, Kuwait. http://www.greenline.com.kw/Reports/047.asp
15. Mubako ST, Lant CL (2013) Agricultural virtual water trade and water footprint of US states. Ann Assoc Am Geogr 103(2):385–396
16. Murdock HE, Collier U, Adib R, Hawila D, Bianco E, Muller S, Ferroukhi R, Renner M, Nagpal D, Lins C (2018) Renewable energy policies in a time of transition
17. Negm AM (2019) Conventional water resources and agriculture in Egypt. Springer
18. Noaman M (2017) Country profile. In: Irrigated agriculture in Egypt. Springer, pp 1–8
19. Okoth-Owiro A (2004) State succession and international treaty commitments: a case study of the Nile water treaties. Occasional paper, East Africa# 9, Konrad Adenauer Stiftung and Law & Policy Foundation
20. Osman R, Ferrari E, McDonald S (2016) Water scarcity and irrigation efficiency in Egypt. Water Econ Policy 2(04):1650009
21. Page EA (2008) Distributing the burdens of climate change. Environ Polit 17(4):556–575
22. Popp J, Kovács S, Oláh J, Divéki Z, Balázs E (2021) Bioeconomy: biomass and biomass-based energy supply and demand. New Biotechnol 60:76–84
23. Shamrukh M, Abdel-Wahab A (2011) Water pollution and riverbank filtration for water supply along river Nile, Egypt. In: Riverbank filtration for water security in desert countries. Springer, pp 5–28
24. Shetty S (2006) Water, food security and agricultural policy in the Middle East and North Africa region. World Bank
25. Snyder RL (1992) Equation for evaporation pan to evapotranspiration conversions. J Irrig Drain Eng 118(6):977–980

26. Tompa O, Lakner Z, Oláh J, Popp J, Kiss A (2020) Is the sustainable choice a healthy choice?—water footprint consequence of changing dietary patterns. Nutrients 12(9):2578
27. Van der Gulik T, Nyvall J (2001) Crop coefficients for use in irrigation scheduling. Ministry of Agriculture, Food and Fisheries of British Columbia. Agdex, 561
28. Whittington D, McClelland E (1992) Opportunities for regional and international cooperation in the Nile basin. Water Int 17(3):144–154
29. Woertz E (2017) Agriculture and development in the wake of the Arab spring. In: Combining economic and political development. Brill Nijhoff, pp 144–169

Cereal Water Footprint in Arid and Semi-arid Regions: Past, Today and Future

Hadi Ramezani Etedali, Mojgan Ahmadi, and Mohammad Bijankhan

Abstract Food security, drought, environmental protection and industrial development have necessitated more efficient management of water resources. Population growth and unsustainable agricultural development in different parts of the world, especially in arid and semi-arid regions, has led to a more realistic approach and the use of a comprehensive and efficient water footprint (WF) index (separately for green water, blue water and gray water) in determining the amount of water consumed by agricultural products. On the other hand, any climate change will cause a change in rainfall patterns and consequently change the share of blue and green water used in agricultural products. Therefore, the importance of cereal WF in arid and semi-arid regions has been discussed. Estimating the ecological WF and virtual water trade in various products in arid and semi-arid regions such as Iran can help better manage the limited water resources. Finally, the importance of accurate water measurement is highlighted to reliably estimate the water footprint.

Keywords Water footprint · Cereal · Water resources · Climate change · Irrigated land · Rainfed land

1 Introduction

1.1 Water Resources Management

Water is one of the challenges of the twenty-first century that can be the source of many positive and negative changes in the world. A mismatch between the supply and the demand could be creating a crisis. This crisis can occur locally, regionally,

H. R. Etedali (✉) · M. Ahmadi · M. Bijankhan
Department of Water Sciences and Engineering, Imam Khomeini International University (IKIU), Qazvin, Iran
e-mail: Ramezani@eng.ikiu.ac.ir

© The Author(s), under exclusive license to Springer Nature Singapore Pte Ltd. 2022
S. S. Muthu (ed.), *Environmental Footprints of Crops*,
Environmental Footprints and Eco-design of Products and Processes,
https://doi.org/10.1007/978-981-19-0534-6_3

nationally or even globally. Imbalances in the water sector can be due to the hydrological cycle and natural limitations of water resources, as well as human activities such as improper use of resources and pollution of resources [1].

Studies show that in 1950, 12 countries in the world with a population of about 20 million people were facing water shortages. Forty years later, that number has risen to 26 countries with a population of about 300 million, and it is predicted that by 2050, 65 countries with a population of more than 7 billion people will face water shortages. There are currently about 840 million people in the world living in food shortages, the majority of which is about 800 million people living in developing countries [2].

Freshwater resources have significant temporal and spatial variations. Population growth combined with socio-economic development has put these resources at risk. Declining groundwater levels, drying up of rivers and high levels of water pollution are signs of water scarcity [3–6].

Agreements between the WWF and international scientific committees have argued for years that water security can no longer be referred to as something unrelated to the value of ecosystem biodiversity as its source. Almost all humans live near water sources. We need water to survive, farm, generate electricity and produce goods for our own consumption. Less than 1% of the world's available water resources supply human needs and the environment stems (two are inseparable) there. The key to ensuring adequate and quality water for humans is to prevent the destruction of water resources such as rivers, lakes and groundwater. Today, the services provided to the economy and human communities by freshwater ecosystems include water supply by extracting water beyond their sustainable level. This is well illustrated in the Millennium Ecosystem Assessment in an international study published in 2005 with the support of the United Nations on the health of planetary ecosystems and their future [7].

In addition, the need for water resources, which is the footprint of human water, is predicted to increase in many parts of the world. The main effects of human water footprint on freshwater ecosystems are seen in increasing river tributaries, extra water extraction and pollution. Sudden current climate change could exacerbate these effects [8].

The growing need for water and hydroelectric energy is seen in efforts to control floods and improve shipping to build dams and other structures such as underground dams and embankments in most of the world's major rivers.

Out of a total of 177 rivers with a length of more than a thousand kilometers, only 64 of them flow without dams or other obstacles. These are in the WWF study published in 2006. Water infrastructure has its advantages, but it can have profound effects on the freshwater ecosystems and populations that are based on these ecosystems. Dams replace river flow, changing the quantity and quality of flow downstream. In addition, larger dams can diversify the entire ecological connection between upstream and downstream habitats and create many issues such as fish migration. Recent studies have shown that the construction of the dam has had a negative impact on the lives of more than 500 million people. Economic and financial crises affect many countries around the world [7]. The chapter roadmap is shown in Fig. 1.

1.2 Importance of Cereals for Food Security

The subject of food and the food component has always been of vital importance in human history and is one of the obvious and natural human rights. Food security is one of the most important aspects of national security, meaning that supply at an optimal level of national security requires the provision of an optimal level of food security [9].

Wheat is one of the most important grains that is considered the dominant force of the people of the world. It is an important source of carbohydrates. Consumption of whole wheat provides nutrients and dietary fiber. Unlike other cereals, wheat can be used in the preparation of bread, biscuits, sweets, cakes, spaghetti, pasta etc.

Corn is a fodder of important products considered as animal feed, especially cattle and sheep. Therefore, the quantity (amount of forage) and quality (percentage of protein, starch, minerals) of its forage are of special importance. This plant is one of the best plants for producing green fodder. Forage corn is a very palatable fodder for cattle and sheep and accepts mechanization conditions well. This type of corn has large amounts of minerals, especially nitrogen, phosphorus, potassium and calcium, is easily digested and its silage is the most suitable feed for livestock.

Food supply is the first and most important determining factor in measuring food security, and therefore, the importance of increasing agricultural and food production, as the most important goal of the agricultural sector, in providing food security for society is evident, although other goals of the agricultural sector such as reduce food waste and increasing farmers' income also play a role in establishing food security in terms of food supply and physical and economic access to the required food [9].

Countries make decisions to improve their food security in accordance with their domestic and foreign relations. This may be achieved through efforts to become self-sufficient in agricultural products or through a combination of domestic production and food imports [9].

In today's world, there is a lot of disruption in the agricultural sector, and food security is not provided to a large part of the population in developing countries.

This is happening in developing countries due to poor infrastructure, including poor storage and packaging of food, rising prices and decline in access to healthy food.

Most of the world's wheat is grown in rainfed conditions. Wheat grown under irrigation conditions requires more cost and water to produce. It will be said that irrigation provides greater security in hot seasons, but if it is dry, irrigation will be a problem, and water in rivers and dams will decrease. Special crops, such as wheat and barley, grow in open fields and require a lot of water, drip irrigation is impractical, and the use of large centrifugal sprinklers in them is unusual and causes

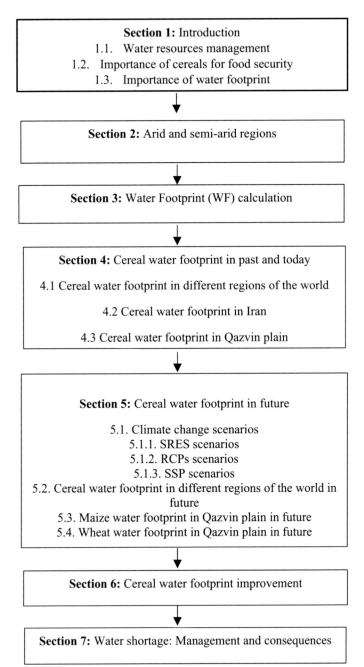

Fig. 1 Chapter roadmap

a large amount of water to be wasted before it reaches the plant. These products use rainwater and rainfed for growth. Bread and other foods made from wheat and barley in rainfed conditions will be cheaper than those produced under irrigation conditions. There is another distinction between wheat as a unique crop and wheat and barley in their growth cycle in Medicago. Wheat grows without this rotation and requires less nitrogen fertilizer and less water is used in the production process [7].

1.3   Importance of Water Footprint

In recent years, a new framework called water footprint (WF) is suggested by Hoekstra [10], which is similar to the ecological footprint [11–13]. This concept, reviewed by Hoekstra and Chapagin [14], makes it possible to analyze the relationship between consumption and the allocation of freshwater resources.

Water footprint (WF) refers to the part of water supplied from surface or groundwater sources and is used in the production of the product. In contrast, green WF is related to the share of rain [15]. Gray WF of a product is defined as the volume of freshwater for dilution of pollutants that are generated in the manufacturing process of the product. This volume of water is determined based on water quality standards.

The concept of virtual water footprint (total water consumption for the production of a unit of agricultural and non-agricultural goods) has been used at regional, national and international levels to analyze water use efficiency [16]. Due to the consumption of about 90% of water resources for agricultural purposes, the concept of virtual water footprint of agriculture is very important. Studying water footprint in the agricultural sector is of great help to better understand the current situation and improve water resources, especially in arid and semi-arid regions [17].

Many studies have been on virtual water footprint and virtual water trade at various levels of regional, national and international [11, 14, 16, 18–33].

Examining the different components of virtual water footprint and determining each of the elements in the amount of virtual water trade in the agricultural sector are of great help to understand the current situation and improve water resources management in the agricultural sector, especially in areas facing water crisis.

2   Arid and Semi-arid Regions

Arid and semi-arid regions (Fig. 2) house approximately 2.5 billion people and occupy 41% of the earth's surface [34].

In arid regions, evapotranspiration is higher than precipitation. In other words, from a hydrological point of view, the water balance in these environments is negative because due to high temperature and dry air, evaporation and transpiration from the soil and plant surface increase from rainfall. Due to low rainfall, vegetation in these areas is poor and scattered, and wildlife and plants are close to small sources of water.

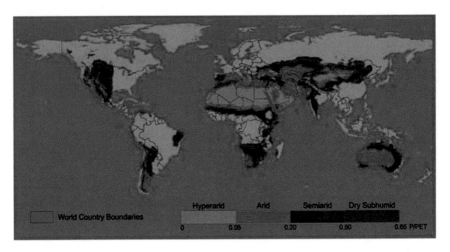

Fig. 2 Global distribution of arid and semi-arid regions (*Source* [35])

Rainfall in these areas is very irregular and sometimes heavy, which is sufficient due to lack of vegetation. And throughout the management of the reservoir, it causes devastating floods and even exists in the vicinity of seas and surrounding areas.

Lack of freshwater resources is one of the most important challenges in many countries. Of course, this is much more the case in countries that are in arid and semi-arid regions and have limited rainfall or high evapotranspiration. Excessive demand for water resources in the face of low rainfall has created conditions that put too much pressure on the ecosystems of these areas. It should be noted that in many countries, especially developing countries today, the watersheds of these areas, which are sources of water production in some parts, have undergone many changes over the past few decades. Reducing the ecological potential of watersheds, reducing the quantity and quality of plant ecosystems, especially rangelands, increasing the amount of erosion and sediment delivery from watersheds and the consequent irreparable damage to water, soil and dams, uncontrolled abstraction of groundwater resources and reducing groundwater aquifers are among the bitter experiences that some watersheds in these areas have experienced over the past few decades [36].

Iran is located in the arid and semi-arid belt of the world and is far from moisture sources. Therefore, rain reaches Iran when they have lost a large part of their moisture and do not have enough capacity to generate rainfall in the central and eastern parts of Iran. In addition, the Zagros and Alborz mountain ranges, although they strengthen the rainfall systems entering the country, by being in the path of air currents, cause the erosion of rainy air masses before reaching the central and eastern regions of the country. Irregularity and severe fluctuation of rainfall from year to year have caused the occurrence of rainfall in Iran, especially in the central and eastern half of the country, completely random and irregular. These irregularities are the main cause of severe and long-term droughts in the country, which in most cases cover a large area of the country and cause irreparable damage to the country's economic structure, especially the agricultural sector [37].

Qazvin province is located in the central part of Iran with an area of 15,821 km², between 48° and 45 min to 50° and 50 min east longitude and 35° and 37 min to 36° and 45 min north latitude (Fig. 2). Qazvin plain is one of the plains of the salt lake catchment and its largest plain has the highest area under cultivation of various crops among the plains of this catchment. It has a semi-arid climate with relatively hot summers and relatively cold winters.

Qazvin plain is one of the most important agricultural hubs of Iran. Due to the recent drought and the increase in the share of various sectors such as industry, environment and drinking, the agricultural sector is facing a severe shortage of water resources [38, 39]. Farmers have increased groundwater abstraction due to the declining share of surface water resources. Improper abstraction of groundwater for irrigation of agricultural lands has caused a sharp drop in the water level in the aquifer. The crisis in the Qazvin plain has made water management in general [40] and in agriculture and cropping pattern reform in particular necessary [41] (Fig. 3).

Due to population growth, the need for food supply, environmental protection and sustainable management of surface and groundwater resources, virtual water footprint and virtual water trade are considered dynamic concepts for water resources management in all sectors that have been considered in recent years [3–6, 42–44].

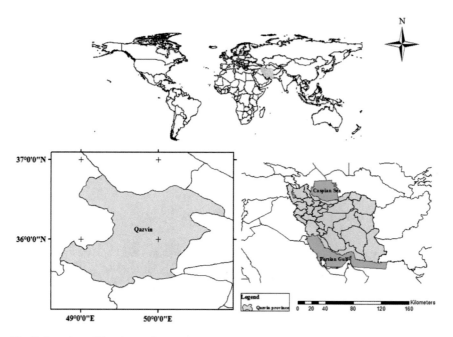

Fig. 3 Location of Qazvin province

3   Water Footprint (WF) Calculation

Water footprint (WF) is an indicator of the volume of water that is used directly or indirectly to produce goods. The blue WF refers to the volume of water that has been sourced from surface or groundwater resources and has either evaporated. Irrigated agriculture, industry and domestic water use can each have a blue water footprint. The green WF refers to the amount of water from precipitation. The gray WF refers to the volume of freshwater required to dilute the fertilizers and toxins used in the production process [15, 45]. The white WF is pertinent to irrigation water loss and is dependent on irrigation efficiency [46]. The WF components were determined by the following equations:

$$WF_{Green} = \frac{P_e \times 10}{Y} \tag{1}$$

$$WF_{Blue} = \frac{(ET_c - P_e) \times 10}{Y} \tag{2}$$

$$WF_{Gray} = \frac{\alpha \times NAR}{C_{Max} - C_{Nat}} \times \frac{1}{Y} \tag{3}$$

$$WF_{White} = \frac{10 \times (D_t - (ET_c - P_e))}{Y} \tag{4}$$

in which WF_{Green} is the green WF, WF_{Blue} the blue WF, WF_{Gray} the gray WF and WF_{White} the white WF in $m^3 \; t^{-1}$. P_e is the total effective rainfall during the crop-growing season (mm), ET_c the crop evapotranspiration (mm), Y the crop yield ($t \; ha^{-1}$), α the percentage of nitrogen fertilizer loss, NAR the rate of fertilizer application ($kg \; ha^{-1}$), C_{Max} the nitrogen critical concentration ($kg \; m^{-3}$), C_{Nat} is the real nitrogen concentration in the receiving water ($kg \; m^{-3}$), D_t is the irrigation depth during the growing season (mm), and "10" is a conversion factor from mm to $m^3 \; ha^{-1}$.

4   Cereal Water Footprint in Past and Today

Examining the different parts of virtual water footprint (WF) and determining the share of each part in the agricultural sector are of great help to better understand the current situation and improve existing water resources, especially in the arid and semi-arid regions.

In order to properly evaluate water consumption in the agricultural sector, it is necessary to study the WF index in different climates [47].

WF assessment is versatile and can inform a broad range of strategic actions and policies from environmental, social and economic perspectives. Many studies have been performed on water footprint [45, 48–54].

4.1 Cereal Water Footprint in Different Regions of the World

In a study in Mexico on wheat, due to low yield and water use efficiency, the amount of irrigated water was estimated to be about 1250–625 (m^3 ton^{-1}). On the other hand, except green due to the climatic conditions in which wheat grows was low, the gray footprint of wheat was 19,364 (m^3 ton^{-1}). It was stated that water footprint not only assesses the use of resources but also the impacts and shortcomings that lead to such uses. A method with these characteristics can provide more appropriate information for decision-making for both scientific research and the general public [55].

Schyns et al. [43] estimated the volume of water footprint components in the production of crops in Jordan for green, blue and gray water at 493, 406 and 54.3 million cubic meters (MCM) per year, respectively. They reported the amount of surface water and groundwater in the agricultural water footprint of the agricultural sector as 143 and 263 million cubic meters (MCM), respectively.

Zhau et al. [56] examined the water footprint components in the Yellow River Basin of China during the period 1961–2009. Their results showed that due to increased yield improvement, water and green water footprints in agricultural production have decreased and gray water footprints have increased due to increased consumption of nitrate and phosphate fertilizers.

In China, wheat and maize water footprints were evaluated from 1956 to 2015 and showed that the total water footprint of wheat was 1,580 (m^3 ton^{-1}) and the share of green, blue and gray water footprints was 52, 29 and 19%, respectively. For maize, the total water footprint was 1275 (m^3 ton^{-1}) and green, blue and gray footprints accounted for 73, 3 and 24% of the total water footprint, respectively. In their study, it was stated that most of the total water footprint is obtained from green water, especially for maize production, and this shows that more attention should be paid to rain management in the future. In general, 19 and 24% of the total water footprint is needed to eliminate agricultural water pollution for wheat and maize, which is much higher than the global average, indicating that fertilizer use efficiency should improve in the future [57].

Pahlow et al. [58] showed that South Africa exported about 22% of its total WF of agricultural production between 1996 and 2005. During those years 10,867, 532 and 1089 MCM/year left the country as green, blue and gray WF, respectively.

4.2 Cereal Water Footprint in Iran

According to the data provided by the Ministry of Jihad Agriculture at the provincial level, 12 provinces (for maize crop) and 15 provinces (for wheat and barley crops) were considered as the main cereal-producing provinces.

Information on 15 selected provinces for the main cereal production in Iran is presented (Figs. 4, 5 and 6). These 15 provinces produce a total of 84.3% of irrigated wheat and 91% of rainfed wheat in the country. The selected provinces produce 87.4% of irrigated barley and 79.2% of rainfed barley (a total of 84.7% of the country's total volume). The selected provinces accounted for 95.8% of the national maize productions.

Nitrogen application rates (NARs) in main cereal-producing provinces in irrigated and rainfed lands for the period of 2006–2012 are presented in Fig. 7.

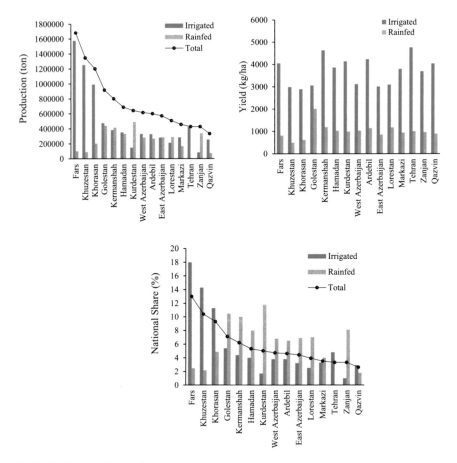

Fig. 4 Production of wheat for selected provinces

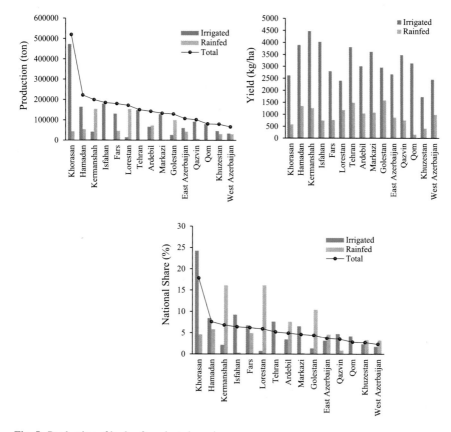

Fig. 5 Production of barley for selected provinces

Wheat

The water footprint (WF) components of wheat production are summarized in Fig. 8. The average total WF in irrigated lands is about 3188 m³/ton, the share of green and blue water is almost equal. Also, the share of gray and white water in irrigated lands in the WF in wheat production is about 52%. Although this amount of water goes back to the country's water cycle, it is possible to apply more effective management to reduce this share. In rainfed lands, the component of green WF varies between 128 and 4166 m³/ton and the gray WF between 100 and 740 m³/ton. In these lands, the average total WF is estimated at 3071 m³/ton, in which the share of green water is nine times the share of gray water. Also, the total WF in irrigated wheat is about 3.7% higher than the total WF in rainfed wheat, which indicates a small difference between the WF in the production of irrigated and rainfed wheat in the country. In other words, in terms of the total water footprint in the production of irrigated wheat and rainfed, there is not much difference. However, in different provinces, in terms of water footprint in irrigated and rainfed lands, there are many differences that can

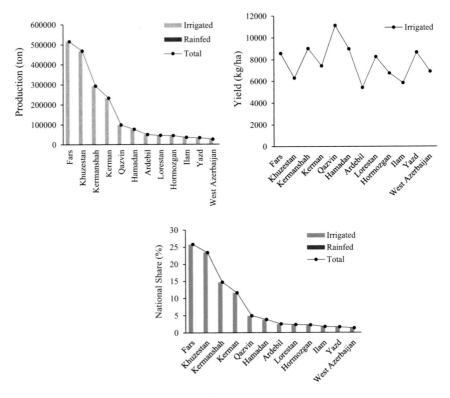

Fig. 6 Production of maize for selected provinces

be used to determine the type of irrigated or rainfed cultivation at the level of these provinces.

Figure 9 shows the national average of the water footprint (WF) components of wheat production. Total WF in irrigated wheat production in three provinces of Golestan, Khorasan and Lorestan is higher than rainfed wheat. Therefore, in these provinces, there can be a change in attitude in choosing the type of wheat cultivation from irrigated to rainfed. In this province, rainfed wheat has no preference over irrigated wheat. In irrigated lands, the share of white WF is 35%. In irrigated lands, Khorasan, Kurdistan and Lorestan provinces have the highest white WF with 46, 40 and 39%, respectively. Therefore, in these provinces, more effective management should be done to control irrigation losses. The lowest white WF is observed in Golestan, Tehran and Qazvin provinces with 29, 31 and 31% of the total WF, respectively. The average share of gray WF in irrigated lands is close to 17%. From this point of view, Ardabil, Fars and Khuzestan provinces have the highest share of gray water with 24, 23 and 23%, respectively, compared to the total WF in each province. In rainfed lands, the share of gray water is 10% and Golestan, Khuzestan and Zanjan provinces with 20, 15 and 15% have the highest share of gray WF compared to the total WF in the province. In irrigated lands, three provinces of Fars, Khorasan and

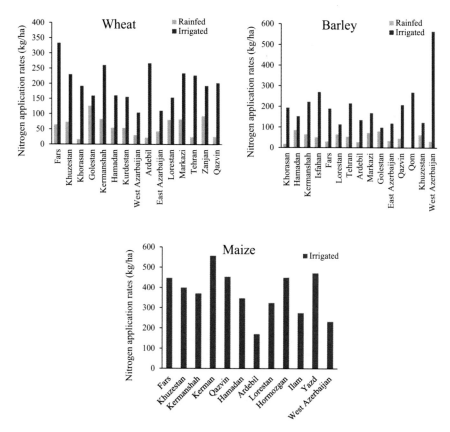

Fig. 7 Nitrogen application rate for main cereals

Khuzestan with 5575, 5028 and 4123 MCM/year, respectively, have the highest water footprint in the country's wheat production. The volume of water in the footsteps of water in wheat production is worrying for those provinces that face water shortages.

Barley

Figure 10 shows the water footprint (WF) components of barley in 15 selected provinces. In the study period, 80.8% of the country's barley cultivation area (85.4% of all irrigated lands and 77.6% of the total rainfed lands) are located in these provinces. Among 15 selected provinces and in lands under irrigated water, the highest and lowest volumes of green water footprints with 1173 and 302 m^3/ton are observed in Lorestan and Isfahan provinces, respectively. Also, the volume of water footprint of 1305 and 435 m^3/ton, respectively, is related to Khorasan and Kermanshah provinces. The volume of gray water footprint varies between 393 m^3/ton in Hamedan province and 2298 m^3/ton in West Azerbaijan province and the volume of white water footprint between 606 m^3/ton in Kermanshah province and 2142 m^3/ton in Khorasan province.

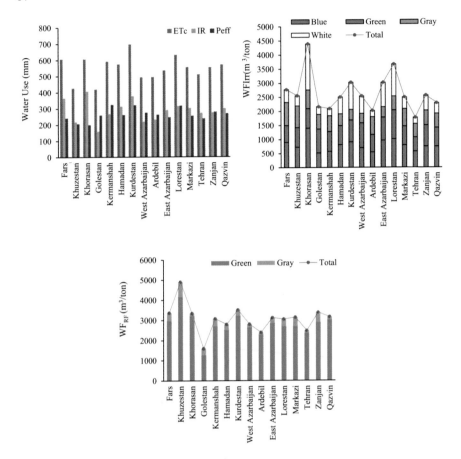

Fig. 8 Wheat water footprint for selected provinces

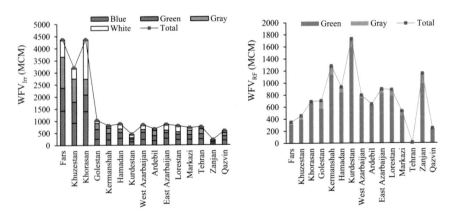

Fig. 9 Total volumes (WFV) of wheat water footprint for selected provinces

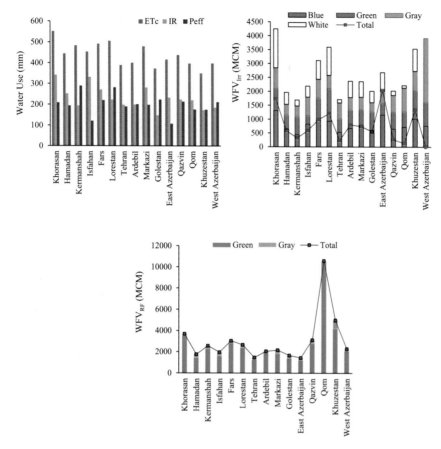

Fig. 10 Barley water footprint for selected provinces

Figure 11 shows the national average of the water footprint (WF) components of barley production. The total WF in the barley lands in the selected provinces varies between 2081 m³/ton in Kermanshah province and 4984 m³/ton in Khorasan province. The average total WF in the production of barley in selected provinces is equal to 3209 m³/ton. Also, the share of green, blue, gray and white water in irrigated lands in selected provinces was estimated at 18, 26, 20 and 36%, respectively. A noteworthy point in the country is the high share of white WF in the production of barley, although this volume of water returns to the country's water cycle and can be used in later years, but with effective management and reducing this share, it can be used in other production processes. Khorasan, Isfahan, Lorestan and Ardabil provinces with 43, 37, 37 and 37%, respectively, have the largest share of white WF (of the total water footprint in each province). This value highlights the importance of controlling irrigation losses and the need for more effective irrigation water management. The lowest share of white WF is related to the provinces of West Azerbaijan, Qazvin,

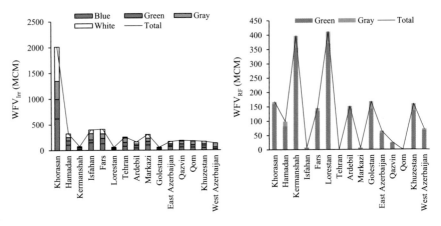

Fig. 11 Total volumes (WFV) of barley water footprint for selected provinces

Tehran and Kermanshah with 21, 29, 29 and 29%, respectively. The highest share of gray water is observed in the provinces of West Azerbaijan, Qom and Tehran with 46, 28 and 23%, respectively.

Maize

Figure 12 shows the national average of the water footprint (WF) components of maize production. In irrigated lands, the minimum and maximum green WF of maize production was 5 and 130 m^3/ton, the blue WF was 567 and 1,172 m^3/ton, the gray WF was 317 and 669 m^3/ton, and the white WF was 181 and 1,302 m^3/ton, respectively. The average total WF for irrigated maize among all the selected provinces was 1,958 m^3/ton, i.e. the highest water productivity among all the three studied crops.

Figure 13 shows the national average of the WF components of maize production. The NTWF of maize production was estimated at around 3,744 MCM/year of which 95.9% of the maize NTWF was related to the main maize-producing provinces.

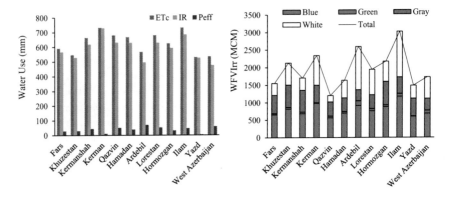

Fig. 12 Maize water footprint for selected provinces

Fig. 13 Total volumes (WFV) of maize water footprint for selected provinces

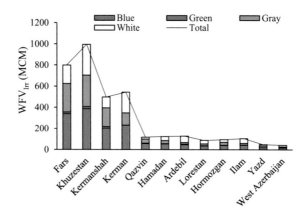

Hoekstra and Chapagain [23] stated that the water footprint of Iran is relatively high and is 1624 (m³/year), partly due to low yield and partly due to high transpiration evaporation due to arid and semi-arid climatic conditions of the country. Also, various studies have shown that wheat has more traces of whole water and green water than other crops. Farzam [59] estimated the wheat water footprint for Iran at 1235 (m³ ton$^{-1 \, year-1}$).

4.3 Cereal Water Footprint in Qazvin Plain

In the Qazvin plain, the area under cultivation of wheat, barley, corn, maize, alfalfa, tomato and canola rapeseed is about 90% of irrigated lands and their products are about 81% of irrigated agricultural production. Also, four crops of wheat, barley, lentils and chickpeas, accounting for about 99% of the area of rainfed lands, produce 92% of the rainfed agricultural products of the region. Wheat with a total of about 144 thousand hectares and production of 315 thousand tons is the most important crop in the cultivation pattern of the region [60].

The average yield, evapotranspiration, amount of applied fertilizer, effective rainfall and irrigation depth for irrigation and rainfed crops during the period of 2003–2014 are shown in Fig. 14 [41, 60].

Due to the high share of white water footprint compared to gray water in the production of water products in the region, dilution of wasted fertilizers will be done by white water footprint and it is not necessary to add more water for this purpose to the total water footprint in the production of water products in the region. Therefore, the total water footprint in the production of water products in the region was considered as the total green, blue and white water footprint. Regarding rainfed cultivation, due to the lack of white water footprint in rainfed lands, gray water footprint was considered as water footprint in the production of rainfed products.

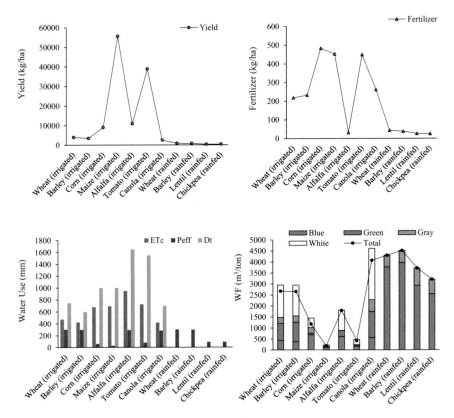

Fig. 14 Water footprint of main rainfed and irrigated crops

The total volume of water footprint in irrigated and rainfed products of the main products was estimated at about 2053 MCM/year. As previously mentioned, due to the high volume of white water footprint in irrigated lands, gray water footprint in the production of water products was omitted. The total volume of water footprint in the production of irrigated and mainland rainfed products of the region is about 1719 and 334 MCM/year, i.e. 84 and 16% of the total water footprint in the production of agricultural products in the region. The share of irrigation losses in the region is about 846 MCM/year, i.e. about 42% of the total volume of water footprint and about 50% of the total volume of water footprint in the production of water products in the Qazvin plain. Also, the volume of gray water footprint is about 43 MCM/year (about 2% of the total volume of water footprint and 53% of the total volume of water footprint in the production of rainfed products). Due to the policy of reducing the share of agricultural water from surface water resources, most of the water required for this sector is supplied from groundwater sources in the region, which ultimately leads to uncontrolled abstraction and severe reduction of groundwater level in the region [61] (Fig. 15).

Fig. 15 Water footprint volumes for main irrigated and rainfed crops

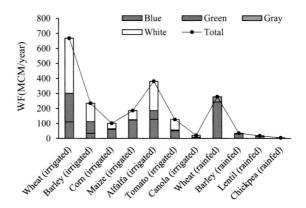

In irrigated lands, about 50% of the total volume of WF in the production of important products is the share of white WF volume. The share of irrigation losses in the region is about 864 MCM/year, i.e. about 42% of the total volume of WF in the Qazvin plain, which is significant for a region facing a water crisis. Also, the share of gray WF, which is the share of the environment in rainfed lands, is about 13%. In total, the volume of gray WFs is about 43 MCM/year (about 2%).

5 Cereal Water Footprint in the Future

One of the methods of modeling climate change is based on the use of mathematical climatic simulations known as general circulation models (GCMs). These models are numerical that simulate our processes between the atmosphere, ocean, ice and surface in three dimensions.

These models take into account many components of the atmosphere and surface properties, such as carbon dioxide, particulate matter and vegetation, and provide estimates of future meteorological parameters [62].

5.1 Climate Change Scenarios

In 1992, the first Intergovernmental Panel on Climate Change (IPCC) release scenarios, called IS 92, were developed for use in the input of general circulation models to model climate change scenarios.

IS2 scenarios include population estimates, energy consumption by trade, industry, transportation and housing, energy production, production and consumption of secondary fuels, energy production from liquid, solid, hydrogen gas, solar cores biomass, carbon dioxide emissions, carbon monoxide, nitro oxides, nitrogen oxides, methane through combustion, methane emissions from mines and many sources of greenhouse gas emissions for ten regions of the globe, including the Americas, Western Europe and Canada, Asia and South Asia Eastern, Central Europe, Central Asia, Africa, the Middle East, Latin America, South and Southwest Asia, and Russia for the period 1985 to 2050 five years apart.

5.1.1 SRES Scenarios

The IPCC published a new set of scenarios in 2000 for use in the Third Assessment Report (Special Report on Emissions Scenarios—SRES). The family of SRES scenarios refers to scenarios that have a common theme and context. In the third and fourth IPCC reports, four families of SRES scenarios are discussed: A1, A2, B1 and B2 [63]. According to IPCC Third and Fourth Reports, the definitions of each release scenario are as follows:

A1: In this scenario, the world is considered integrated. The characteristics of the A1 family are as follows: 1. Rapid economic growth, 2. World population will reach 9 billion in 2050 and then gradually decrease, 3. Rapid expansion of new and efficient technologies, 4. Convergent world, income and lifestyles converge between regions, 5. The rapid expansion of social and cultural interactions in the world.

A2: The characteristics of Scenario A2, in which, unlike Scenario A1, the world is not considered convergent, are as follows: 1. A world in which countries operate independently and are self-reliant; 2. Economic development is region-oriented.

B1: In this scenario, the world is considered integrated and environmentally friendly, the characteristics of which are: 1. Rapid economic growth as in Scenario A1 and based on the provision of services and information; 2. The world population will reach 9 billion by 2050, but then it is similar to Scenario A1; 3. The use of clean resources and new technologies with high efficiency and reduction of pollutants; 4. Emphasis on global solutions for economic, social and environmental sustainability.

B2: In Scenario B2, the world is divergent as in Scenario A2, except that it is considered environmentally friendly. The characteristics of this scenario are as follows: 1. The population is constantly increasing, but its growth rate is slower than A2; 2. Emphasis on current local solutions considering global solutions for economic, social and environmental stability; 3. Economic development is moderate; 4. Technological changes are fast but will be less and more scattered than A1 and B1.

5.1.2 RCPs Scenarios

The IPCC Fifth Assessment Report (AR5) is due for publication in 2013–14. Its findings will be based on a new set of scenarios that replace the SRES standards employed in two previous reports. The new scenarios are called Representative Concentration Pathways (RCPs). There are four RCPs scenarios based on multi-gas emission scenarios, namely RCP2.6, RCP4.5, RCP6 and RCP8.5.

RCP2.6 This scenario involves the lowest rate of increase in greenhouse gases and induction of radiation. According to this scenario, the radiation induction in the middle of this century (between 2030 and 2050) reached about 3 W m^{-2} and then decreased in 2100. To reach this level of radiation induction, greenhouse gases must be significantly reduced.

RCP4.5 and **RCP 6.0** are known as intermediate stabilization pathways scenarios where energy will reach 4.5 W m^{-2} and 6.0 W m^{-2}, respectively, after 2100.

RCP8.5 In this scenario energy will reach 8.5 W m^{-2} by 2100 and continue to rise for some amount of time.

5.1.3 SSP Scenarios

The new scenarios represent different socio-economic developments as well as different pathways of atmospheric greenhouse gas concentrations. SSP scenarios are described based on the five fundamental approaches of sustainable development, regional competition, inequality, fossil fuel development and intermediate policy development. These scenarios fall into five categories known as SSP1 to SSP5.

SSP1 includes sustainable consumption, low population growth, increased energy efficiency, faster renewable energy substitution and greater global cooperation.

SSP2 stands for interface conditions; in these conditions, socio-economic development is in line with normal conditions.

SSP3 envisions a world with high challenges for adjustment-related policies, including high population growth leading to high food and energy demand and regional competition. Such conditions hinder social and technological development.

SSP4 reflects a highly unequal world with inequality in economic and political power, leading to increased inequality at home and abroad throughout the twenty-first century. Conflict and unrest are also assumed to increase. Also, the development of advanced technology in various sectors is great and the energy system will be diverse.

SSP5 is an advanced yet fossil-fuel world that uses energetic lifestyles [64].

5.2 Cereal Water Footprint in Different Regions of the World in Future

Yesilkoy and Saylan [65] evaluated and modeled the yield and water footprint of winter wheat with Aqua Crop in the Thrace region of Turkey. They used the RCP8.5 and RCP4.5 scenarios and the HadGEM2-ES model for future periods 2040–2020, 2041–2070 and 2099–2071 to evaluate water performance and footprint. Their results showed that wheat yield increased and water footprint decreased.

Mali et al. [66] quantified the WF of major cereals crops in India. They used RCP4.5 and RCP6.0 and future periods of 2030s and 2050s. Their result showed WF of the cereal crops will change in the range of—3.2 to 6.3% in future periods.

Garofalo et al. [67] evaluated the WF of winter wheat under climate change in Germany and Italy. Their results showed that the WF was 5% lower on average in Italy and 23% in Germany when compared to the baseline.

Kobuliev et al. [68] evaluated the effect of future climate change on the WF of winter wheat in southern Tajikistan for two future periods (2021–2050 and 2051–2080), under three RCPs (RCP2.6, RCP4.5 and RCP8.5). Their results showed that the WF of winter wheat decreased from 3.4 to 2.2%.

Many studies have been done on crop water footprint and climate change [49, 69–81].

5.3 Maize Water Footprint in Qazvin Plain in Future

The water footprint (WF) of maize simulated with the Aqua Crop model in Qazvin plain in the baseline (1986–2015) is presented in Fig. 16.

The average total water footprint (WF) of maize water in the synoptic station of Qazvin plain was about 260 (m³/ton). The share of green WF was 6.78% and the share of blue WF was 93.23%. The share of blue WF compared to green WF is high, which indicates a low rainfall rate and indicates the stability of arid and semi-arid climates in terms of agriculture [82]. Aligholinia et al. [83] evaluated the blue and green WF in the Urmia watershed. They examined the WF of five major crops, including wheat, sugar beet, tomato, alfalfa and maize. Their results showed that the share of green and blue WFs are 25% and 75%, respectively, and the share of water is higher.

Results of maize water footprint (WF) estimated by the Aqua Crop model in Qazvin plain for RCPs scenarios and general circulation models in LARS-WG model (EC-Earth, GFDL-CM3, HadGEM2-ES, MIROC5, MPI-ESM-MR) for the period

Fig. 16 Water footprint of maize in the Qazvin plain

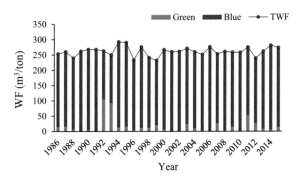

2021–2100 and the percentage of maize water footprint changes compared to the baseline are shown in Figs. 17 and 18.

The minimum water footprint (WF) in the baseline was 232.11 (m³/ton). The minimum WF in the period 2021–2040, 2041–2060, 2061–2080 and 2081–2100 in the model EC-EARTCH were estimated to be 249.79, 248.13, 245.19 and 220.32 (m³/ton), respectively.

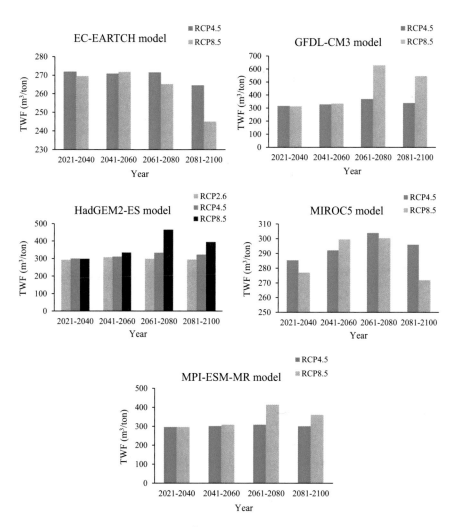

Fig. 17 Average total water footprint (m³/ton) of maize in Qazvin plain

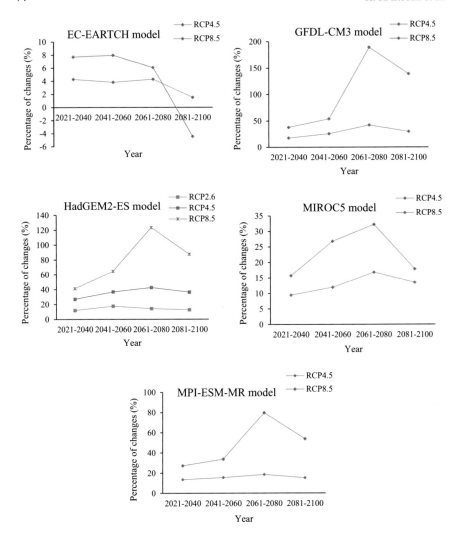

Fig. 18 Percentage of changes maize water footprint in Qazvin plain

5.4 Wheat Water Footprint in Qazvin Plain in Future

The water footprint of wheat simulated with the Aqua Crop model in Qazvin plain in the baseline (1986–2015) is presented in Fig. 19.

The average total water footprint (WF) of wheat water in the synoptic station of Qazvin plain was about 410 (m³/ton). The share of green WF was 91.7% and the share of blue WF was 8.3%. Ababaei and Ramezani Etedali [84] examined the water footprint index for wheat, barley and maize crops for the period 2006 to 2012. Their

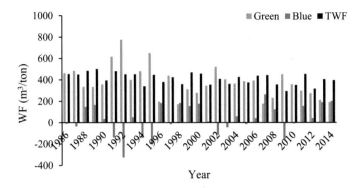

Fig. 19 Water footprint of wheat in Qazvin plain

results showed that the green WF of these crops in the country, for wheat and barley, is 2.3 and 1.9 times more than the blue WF, respectively.

Results of wheat water footprint (WF) estimated by the Aqua Crop model in Qazvin plain for RCPs scenarios and general circulation models in LARS-WG model (EC-Earth, GFDL-CM3, HadGEM2-ES, MIROC5 MPI-ESM-MR) for the period 2021–2100 and the percentage of performance changes compared to the baseline are shown in Figs. 20 and 21.

The minimum water footprint (WF) in the baseline was 294.5 (m^3/ton). The minimum WF estimated during the period 2021–2040 in model HadGEM2-ES is equal to 317.29 (m^3/ton). For the period 2041–2060 in model GFDL-CM3 it is equal to 260.07 (m^3/ton); for 2061–2080 in model HadGEM2-ES it is equal to 192.55 (m^3/ton) and for 2081–2100 in model EC-EARTCH it is equal to 195.91 (m^3/ton).

6 Cereal Water Footprint Improvement

Food security, drought, environmental protection and industrial development have necessitated more efficient management of water resources. The concept of virtual water footprint has significant potential to help improve water management, especially in the agricultural sector. Determination of cropping pattern-based virtual water is a good solution to solve the water crisis, especially in areas with low water and dry climate. Therefore, instead of producing products with high water consumption, it is possible to produce products with less water consumption and reduce the excessive pressure on existing water resources [85].

Estimation of water footprint components in the production process (both in agriculture and industry) can be considered as an important part of water resources management studies. Such studies, by identifying the areas with the highest share of each of the water footprint components, make it possible to more purposefully manage and implement more effective strategies for managing water resources with

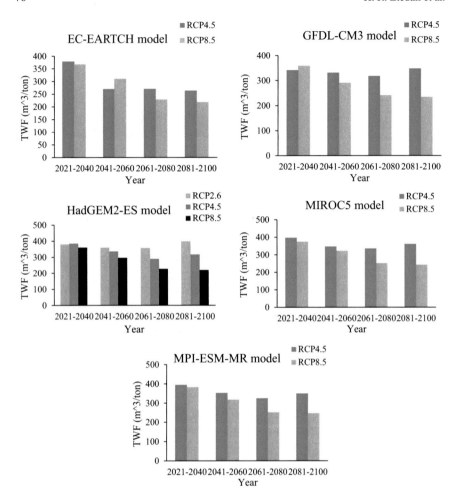

Fig. 20 Average total water footprint of wheat in Qazvin plain

the aim of increasing productivity and reducing economic and environmental costs. Also, in order to more accurately estimate the components of water footprint in the field of crop production (especially cereals), the use of plant models is recommended. These models, taking into account all the factors affecting water requirement and crop yield, provide the possibility of more accurate estimation of these components at the spatial scale and also study the trend of changes in these factors and water footprint components over time and under the influence of different climatic and management scenarios.

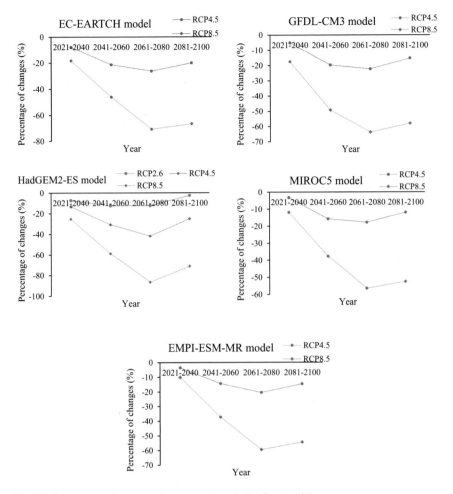

Fig. 21 Percentage of changes wheat water footprint in Qazvin plain

7   Water Shortage: Management and Consequences

As a general definition, water footprint is the ratio of the water use over yield. Hence, accurate flow measurement is a key to calculate reliable water footprint values. On the other hand, effective water management requires a comprehensive view of how much water is being used. Agricultural water needs impose a significant demand. Therefore, the knowledge of precise measurement of the agricultural water use would be a key issue to plan for the available water successfully. Groundwater is one of the main sources of water supply in arid and semi-arid areas. Excessive groundwater depletion results in land subsidence. In this section different methods of the flow measurement of agricultural water wells are presented.

Fig. 22 Well pump outlet **a** full pipe flow, **b** partially filled pipe flow

Different kinds of flow measurement devices could be used for well depletion monitoring. Among all, portable ultrasonic meter is one of the most commonly used devices. However, field conditions are always subjected to many measurement difficulties. Such uncertainties may limit the application of the measurement instruments and their performance.

The following cases were observed based on the field inspections:

1. A full pipe flow (Fig. 22a)
2. Partially filled pipe flow (Fig. 22b)
3. Unsteady outflow condition.

Due to the significant water table decrease during the last decades, in some cases, well discharges were highly reduced; hence, either partially filled pipe flow or unsteady outflow conditions took place. Ultrasonic flow meters could be used only for full pipe flow cases, however, field observations indicated some difficulties, mainly classified in pipe wall thickness variations and limited pipe length to install the sensors.

For partially filled pipe cases, neither magnetic nor ultrasonic flow meters could be used. Employing the basic hydraulic concepts of brink flow conditions could be used. To this end, the flow at the pipe outlet should be measured. Rajaratnam and Muralidhar [86] presented a graphical solution to find the head-discharge relationship of a partially filled pipe (Fig. 22). Equation (5) could also be used to find the flow rate directly:

$$\frac{Q}{g^{0.5}D^{2.5}} = 0.76\left(\frac{y_e}{D}\right)^{0.52} \tag{5}$$

in which Q is the flow through a partially filled pipe, D is the pipe diameter, y_e is the brink depth and g is the acceleration due to gravity. Note that Eq. (5) is only valid for the horizontal pipes (Fig. 23).

Fig. 23 Graphical solution for discharge prediction in partially filled pipes

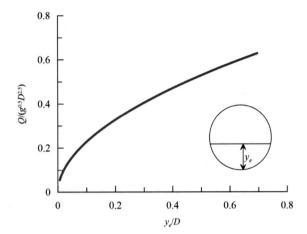

As a conclusion, a reliable water footprint value depends on an accurate water use measurement. In this section a detailed protocol was presented for well depletion measurements.

References

1. Kani Golzar M (2012) Water consumption management based on virtual water exchange in selected products of the country. Master thesis in Agricultural Management, University of Tehran
2. IWMI (2000) World water and climate Atlas. http://WWW.iwmi.org
3. Gleick PH (1993) Water in crisis: a guide to the world's fresh water resources. Oxford University Press, Oxford
4. Mekonnen MM, Hoekstra AY (2010) A global and high-resolution assessment of the green, blue and grey water footprint of wheat. Hydrol Earth Syst Sci 14:1259–1276
5. Postel SL (2000) Entering an era of water scarcity: the challenges ahead. Ecol Appl 10(4):941–948
6. WWAP (2009) The United Nations world water development report 3: water in a changing world, World Water Assessment Programme. UNESCO Publishing, Paris/Earthscan, London
7. Antonelli M, Greco F (2015) The water we eat combining virtual water and water footprints. Springer Water. https://doi.org/10.1007/978-3-319-16393-2
8. WWF (2012) Living planet report 2012, in collaboration with the Global Footprint Network, Zoological Society of London, European Space Agency. www.panda.org
9. Ebadi F, Nasr Isfahani I (2015) Three discourses on concepts Food security, self-sufficiency and extraterritorial agriculture, Planning research, agricultural economics and rural development
10. Hoekstra AY (2003) Virtual water trade. In: Proceedings of the international expert meeting on virtual water trade, Delft, The Netherlands, 12–13 December 2002, Value of Water Research Report Series No.12, UNESCO-IHE, Delft
11. Wackernagel M, Jonathan L (2001) Measuring sustainable development: ecological footprints. Centre for Sustainability Studies, Universidad Anahuac de Xalapa, Mexico
12. Wackernagel M, Rees W (1996) Our ecological footprint: reducing human impact on the Earth. New Society Publishers, Gabriola Island, B.C.

13. Wackernagel M, Onisto L, Linares AC, Falfan ISL, Garcia JM, Guerrero IS, Guerrero MGS (1997) Ecological footprints of nations: how much nature do they use? How much nature do they have? Centre for Sustainability Studies, Universidad Anahuac de Xalapa, Mexico
14. Hoekstra AY, Chapagain AK (2008) Globalization of water: sharing the planet's freshwater resources. Blackwell Publishing, Oxford
15. Hoekstra AY, Chapagain AK, Aldaya MM, Mekonnen MM (2009) Water footprint manual: state of the art 2009. Water Footprint Network, Enschede
16. Allan JA (1997) Virtual water: a long-term solution for water short Middle Eastern economies. Paper presented at the 1997 British Assoc. Festival of Sci., University of Leeds
17. Khalili T, Sarai TM, Babazadeh H, Ramezani EH (2020) Water resources management of Qom Province by using the concept of water footprint. J Ecohydrol 6(4):1109–1119
18. Aldaya MM, Allan JA, Hoekstra AY (2010) Strategic importance of green water in international crop trade. Ecol Econ 69(4):887–894
19. Aldaya MM, Hoekstra AY (2010) The water needed for Italians to eat pasta and pizza. Agric Syst 103:351–360
20. Antonelli M, Sartori Y (2015) Unfolding the potential of the virtual water concept. What is still under debate? Environ Sci Policy 50(2):240–251
21. Faramarzi M, Yang H, Mousavi J, Schulin R, Binder C, Abbaspour K (2010) Modelling blue and green water resources availability in Iran. Hydrol Earth Syst Sci Discuss 7(3):2609–2649
22. Gerbens-Leenes W, Hoekstra AY, Van der Meer TH (2009) The water footprint of bioenergy. Proc Natl Acad Sci 106(25):10219–10223
23. Hoekstra AY, Chapagain AK (2007) Water footprints of nations: water use by people as a function of their consumption pattern. J Water Resour Manag 21(1):35–48
24. Hoekstra AY, Hung PQ (2002) Virtual water trade: a quantification of virtual water flows between nations in relation to international crop trade. Value of Water Research Report Series No. 11, UNESCO-IHE, Delft
25. Hoekstra AY, Hung PQ (2005) Globalisation of water resources: international virtual water flows in relation to crop trade. Glob Environ Changes 15(1):45–56
26. Ababaei B, Ramezani EH (2014) Estimation of water footprint components of Iran's wheat production: comparison of global and national scale estimates. J Environ Process 1:193–205
27. Liu J, Williams JR, Zehnder AJB, Yang H (2007) GEPIC—modeling wheat yield and crop water productivity with high resolution on a global scale. Agric Syst 94:478–493
28. Liu J, Yang H (2010) Spatially explicit assessment of global consumptive water uses in cropland: green and blue water. J Hydrol 384:187–197
29. Liu J, Zehnder AJB, Yang H (2009) Global consumptive water use for crop production: the importance of green water and virtual water. Water Resour Res 45:W05428. https://doi.org/10.1029/2007WR006051
30. Oki T, Kanae S (2004) Virtual water trade and world water resources. Water Sci Technol 49(7):203–209
31. Portmann F, Siebert S, Bauer C, Doll P (2008) Global data set of monthly growing areas of 26 irrigated crops. Frankfurt Hydrology Paper 06, Institute of Physical Geography, University of Frankfurt, Frankfurt am Main
32. Tian G (2013) Effect of consumption of livestock products on water consumption in China based on virtual water theory. Int Conf Future Inf Eng 5(3):112–117
33. Arrien MM, Aldaya MM, Rodriguez CI (2021) Water footprint and virtual water trade of maize in the Province of Buenos Aires, Argentina. Water 13(13):1769. https://doi.org/10.3390/w13131769
34. Gaur MK, Squires VR (2018) Geographic extent and characteristics of the world's arid zones and their peoples. In: Climate variability impacts on land use and livelihoods in drylands. Springer, Cham
35. IIASA/FAO (2003) Compendium of agricultural-environmental indicators (1989–91 to 2000). Statistics Analysis Service, Statistics Division, Food and Agriculture Organization of United Nations, Rome

36. Kousari K, Ekhtesasi MR, Malekinezhad H (2017) Investigation of long term drought trend in semi-arid, arid and hyper-arid regions of the world. Iran Sci Assoc Desert Manag Control 8:36–53

37. Raziei T (2016) Investigation of drought characteristics in arid and semi-arid regions of Iran. Watershed Eng Manag 7(4):363–378

38. Daneshkar Arasteh P, Shokoohi AR (2008) In search of the effects of climate change on weather conditions and surface water resources in Iran. In: 3rd Conference of Iran water resources management, Tabriz

39. Shokoohi AR, Raziei T, Daneshkar AP (2014) On The effects of climate change and global warming on water. Int Bull Water Resour Dev 2(4):1–9

40. Shokoohi AR (2012) Comparison of SPI and RDI in drought analysis in lical scale with emphasizing on agricultural drought (Case study: Qazvin and Takestan). Irrig Water J 3(9):111–122 (in Persian with English abstract)

41. Ramezani Etedali H, Ahmadaali K, Liaghat A, Parsinejad M, Tavakkoli AR, Ababaei B (2015) Optimum water allocation between irrigated and rainfed lands in different climatic conditions. Biol Forum Int J 7(1):1556–1567

42. Norse D (2005) Non-point pollution from crop production: global, regional and national issues. Pedosphere 15(4):499–508

43. Schyns JF, Hamaideh A, Hoekstra AY, Mekonnen MM, Schyns M (2015) Mitigating the risk of extreme water scarcity and dependency: the case of Jordan. Water 7:5705–5730

44. Yang H, Wang L, Abbaspour KC, Zehnder AJ (2006) Virtual water highway: water use efficiency in global food trade. J Hydrol Earth Syst Sci 3(1):1–26

45. Hoekstra AY, Chapagain AK, Aldaya MM, Mekonnen MM (2011) The water footprint assessment manual: setting the global standard. Water Footprint Network, Enschede

46. Ababaei B, Ramezani EH (2016) Water footprint components of cereal production in Iran. Agric Water Manag. https://doi.org/10.1016/j.agwat.2016.07.016. Corrected Proof. Accessed 21 July 2016

47. Aligholi Nia T, Sheibani H, Mohammadi O, Hesam M (2019) Evaluation and comparison of blue, green and gray water footprint of wheat in different climates of Iran. Iran-Water Resour Res 15(3):234–245

48. De Miguel A, Kallache M, García-Calvo E (2015) The water footprint of agriculture in Duero River Basin. Sustainability 7(6):6759–6780

49. Elbeltagi A, Deng J, Wang K, Hong Y (2020) Crop water footprint estimation and modeling using an artificial neural network approach in the Nile Delta. Agric Water Manag 235:106080. https://doi.org/10.1016/j.agwat.2020.106080

50. Karandish F, Simůnek J (2018) An application of the water footprint assessment to optimize production of crops irrigated with saline water: a scenario assessment with HYDRUS. Agric Water Manag 208:67–82

51. Nezamoleslami R, Hosseinian SM (2020) An improved water footprint model of steel production concerning virtual water of personnel: the case of Iran. J Environ Manag 260:110065

52. Dai C, Qin XS, Lu WT (2021) A fuzzy fractional programming model for optimizing water footprint of crop planting and trading in the Hai River Basin, China. J Clean Prod 278:123196

53. Mokarram M, Zarei AR, Etedali HR (2021) Optimal location of yield with the cheapest water footprint of the crop using multiple regression and artificial neural network models in GIS. Theor Appl Climatol 143(1):701–712

54. Rajakal JP, Ng KS, D, Tan T, Andiappan V, Wan YK (2021) Multi-objective expansion analysis for sustainable agro-industrial value chains based on profit, carbon and water footprint. J Clean Prod. https://doi.org/10.1016/j.jclepro.2020.125117

55. Carole F, Sylvie T, Nydia S (2011) Assessment of the water footprint of wheat in Mexico. Towards life cycle management, pp 161–170

56. Zhau L, Mekonnen MM, Hoekstra AY, Wada Y (2016) Inter- and intra-annual variation of water footprint of crops and blue water scarcity in the Yellow River Basin (1961–2009). Adv Water Resour 87:29–41

57. Yuping H, Dongdong J, La Z, Sabine S, JoseMiguel S, Huiping H, Chunying W (2018) Assessing the water footprint of wheat and maize in Haihe River Basin, Northern China (1956–2015). J Water 10:1–18
58. Pahlow M, Snowball J, Fraser G (2015) Water footprint assessment to inform water management and policy making in South Africa. Water SA 41(3):301–305
59. Farzam N (2015) Green and blue waterfootprint of growing crops in Iran and Finland: a case study of six primary crops during 2007–2012. MSc thesis, Helsinki University
60. Agriculture Jihad Ministry (2015) http://www.maj.ir
61. Tehran Regional Water Company (TRWC) (2006) Review of Qazvin irrigation and drainage network. Final report
62. Taylor KE, Stouffer RJ, Meehl GA (2012) An overview of CMIP5 and the experiment design. Bull Am Meteor Soc 93:485–498
63. IPCC (2007) Summary for policy makers climate change: the physical science basis. Contribution of working group I to the forth assessment report. Cambridge University Press, 881 pp
64. Estoque RC, Ooba M, Togawa T, Hijioka Y (2020) Projected land-use changes in the shared socioeconomic pathways: insights and implications. Ambio
65. Yesilkoy S, Saylan L (2020) Assessment and modelling of crop yield and water footprint of winter wheat by aqua crop. Ital J Agrometeorol 3:3–14
66. Mali SS, Shirsath PB, Islam A (2021) A high-resolution assessment of climate change impact on water footprints of cereal production in India. Sci Rep 11:8715. https://doi.org/10.1038/s41598-021-88223-6
67. Garofalo P, Ventrella D, Christian Kersebum K, Gobin A, Trnka M, Giglio L, Dubrovsky M, Castellini M (2019) Water footprint of winter wheat under climate change: trends and uncertainties associated to the ensemble of crop models. Sci Total Environ 658:1186–1208
68. Kobuliev M, Liu T, Kobuliev Z, Chen X, Gulakhmadov A, Bao A (2021) Effect of future climate change on the water footprint of major crops in southern Tajikistan. Reg Sustain 2(1):60–72
69. Elbeltagi A, Rizwan M, Malik A, Mehdinejadiani B, Srivastava A, Singh A, Deng J (2020) The impact of climate changes on the water footprint of wheat and maize production in the Nile Delta, Egypt. Sci Total Environ 743:140770. https://doi.org/10.1016/j.scitotenv.2020.140770
70. Elbeltagi A, Deng J, Wang K, Malik A, Maroufpoor S (2020) Modeling long-term dynamics of crop evapotranspiration using deep learning in a semi-arid environment. Agric Water Manag 241:106334. https://doi.org/10.1016/j.agwat.2020.106334
71. Zheng J, Wang W, Ding Y, Liu G, Xing W, Cao X, Chen D (2020) Assessment of climate change impact on the water footprint in rice production: historical simulation and future projections at two representative rice cropping sites of China. Sci Total Environ 709:136190
72. Deihimfard R, Rahimi-Moghaddam S, Collins B, Azizi K (2021) Future climate change could reduce irrigated and rainfed wheat water footprint in arid environments. Sci Total Environ 150991
73. Govere S, Nyamangara J, Nyakatawa EZ (2020) Climate change signals in the historical water footprint of wheat production in Zimbabwe. Sci Total Environ 742:140473
74. Ahmadi M, Ramezani Etedali H, Elbeltagi A (2021) Evaluation of the effect of climate change on maize water footprint under RCPs scenarios in Qazvin plain, Iran. Agric Water Manag 245:106969
75. Bocchiola D, Nana E, Soncini A (2013) Impact of climate change scenarios on crop yield and water footprint of maize in the Po valley of Italy. Agric Water Manag 116:50–61
76. Arunrat N, Pumijumnong N, Sereenonchai S, Chareonwong U, Wang C (2020) Assessment of climate change impact on rice yield and water footprint of large-scale and individual farming in Thailand. Sci Total Environ 726:137864
77. Shrestha S, Chapagain R, Babel MS (2017) Quantifying the impact of climate change on crop yield and water footprint of rice in the Nam Oon Irrigation Project, Thailand. Sci Total Environ 599:689–699
78. Sun SK, Wu PT, Wang YB, Zhao XN (2012) Impacts of climate change on water footprint of spring wheat production: the case of an irrigation district in China. Span J Agric Res 4:1176–1187

79. Yang M, Xiao W, Zhao Y, Li X, Huang Y, Lu F, Hou B, Li B (2018) Assessment of potential climate change effects on the rice yield and water footprint in the Nanliujiang catchment, China. Sustainability 10(2):242

80. Zhang Y, Huang K, Yu Y, Hu T, Wei J (2015) Impact of climate change and drought regime on water footprint of crop production: the case of Lake Dianchi Basin, China. Nat Hazards 79(1):549–566

81. Zhang Y, Huang K, Yu Y, Wu L (2020) An uncertainty-based multivariate statistical approach to predict crop water footprint under climate change: a case study of Lake Dianchi Basin, China. Nat Hazards 104(1):91–110

82. Aligholinia T, Sheibani H, Mohammadi O, Hesam M (2019) Evaluation and comparison of blue, green and gray water footprint of wheat in different climates of Iran. J Iran Water Resour Res 15(3):234–245

83. Aligholinia T, Rezaie H, Behmanesh J, Montaseri M (2016) Presentation of water footprint concept and its evaluation in Urmia lake watershed agricultural crops. J Water Soil Conserv 23(3):337–344

84. Ababaei B, Ramezani Etedali H (2017) Water footprint assessment of main cereals in Iran. Agric Water Manag 179:401–411

85. Turton AR (2000) Precipitation, people, pipelines and power: towards a virtual water based political ecology discourse. MEWREW occasional paper. Water Issues Study Group. School of Oriental and African Studies (SOAS). University of London

86. Rajaratnam N, Muralidhar D (1964) End depth for circular channels. J Hydraulic Div ASCE 90(2):99–119

Environmental Footprints of Hydrogen from Crops

Alisson Aparecido Vitoriano Julio, Túlio Augusto Zucareli de Souza, Danilo Henrique Donato Rocha, Christian Jeremi Coronado Rodriguez, José Carlos Escobar Palacio, and José Luz Silveira

Abstract The environmental footprint of crops has become one of the main parameters for assessing the ecological impact of food production. However, with the development of renewable technologies for fuel and energy production from biomass-based feedstock, these environmental indicators were also extended to the energy sector. Among the fuels that are likely to soon increase its participation in the energy matrix, hydrogen must receive special attention due to its high energy content and carbon-free combustion. This fuel, however, remains dependent on fossil sources such as natural gas and oil-derived compounds, while production of the so-called "green hydrogen" remains a secondary option. Aiming to understand and quantify the potential decrease of environmental impact by moving toward more renewable hydrogen production pathways, several studies were carried out over the years in order to assess the real impact of this fuel's production through land, water, energy, and other environmental indicators. In this sense, this chapter provides an up-to-date overview of the impact behind hydrogen production, including the three main options currently available: thermochemical processes, biological conversion, and electrolysis. Finally, the main findings allow a deep understanding of potential benefits to be achieved by making the hydrogen matrix more sustainable, while also presenting the main barriers that should be overcome in order for this goal to be achieved.

Keywords Hydrogen · Steam reforming · Biomass · Environmental footprint · Thermochemical conversion · Sustainability · Energy transition

A. A. V. Julio · T. A. Z. de Souza (✉) · D. H. D. Rocha · C. J. C. Rodriguez · J. C. E. Palacio
Institute of Mechanical Engineering, Federal University of Itajubá (UNIFEI), Itajubá, MG, Brazil
e-mail: tulio_zucareli@unifei.edu.br

J. L. Silveira
Waste Revaluation Center, Federal University of ABC (UFABC), São Bernardo do Campo, Brazil

Institute of Bioenergy Research, Sao Paulo State University, IPBEN-UNESP, Guaratinguetá, Brazil

© The Author(s), under exclusive license to Springer Nature Singapore Pte Ltd. 2022
S. S. Muthu (ed.), *Environmental Footprints of Crops*,
Environmental Footprints and Eco-design of Products and Processes,
https://doi.org/10.1007/978-981-19-0534-6_4

1 Introduction

Over the years, the increase in energy demand and environmental regulations led to the intensive search for new fuels and sources of energy. Among them, hydrogen stands out as a trend for the future, mainly due to important particularities such as extremely high energy content and carbon-free combustion [1]. Despite being often presented as a "green" option, hydrogen production is strongly dependent on fossil fuels such as natural gas, in such a way that the absence of carbon in its combustion does not make its life cycle less carbon-intensive [2]. Aiming to tackle this issue and decrease the carbon and other environmental footprints of the hydrogen matrix, several alternative pathways were proposed along the time [3, 4], but up to now, it was not enough to ensure a significant decrease in the role played by fossil fuels in this industry [5].

Currently, hydrogen is mostly produced through thermochemical processes such as steam reforming of natural gas [6], a process that will be further detailed in this chapter. However, alternative technologies have been applied over the years, ranging from electrolysis [7] to biological [8] and other thermochemical processes [9]. In this sense, crops can represent an important feedstock for hydrogen production, since they can replace fossil fuels in both conventional and alternative pathways. In fact, as will be discussed in this chapter, a large variety of crops has been successfully used for producing hydrogen through biological and thermochemical pathways [10].

With the goal to deepen the understanding of the viability of different hydrogen-producing pathways, intensive research has been carried out to assess the existing options from economic, energy, and environmental points of view. Even though economic and logistic perspectives seem to favor the traditional fossil-based processes [11], environmental analysis indicates the strong potential to be explored in alternative pathways, especially regarding processes that use crop-derived feedstock [2]. These studies commonly apply environmental footprints such as land, water, carbon, and more generic footprints as a tool to quantitatively compare different production pathways, creating a wider understanding of the potential improvements to be achieved in the hydrogen industry from an environmental point of view [12].

As will be discussed in this chapter, the use of crops for hydrogen production can be extremely versatile, since several crop-derived products can be used for this end. Actually, crops may be used not only as a feedstock but also as a source of energy for hydrogen-producing pathways that require a heat input [4]. Depending on the technology employed, even crops by-products such as straw [13] and husks [14] can be used to produce this high-quality fuel, showing the strong potential that agriculture can play in the transition toward a cleaner and less fossil-dependent hydrogen matrix.

Besides the increased flexibility of using different pathways for hydrogen production, crops are obviously a renewable source of energy, which may lead to the idea that hydrogen produced from crops and its by-products is more environmentally friendly when compared to traditional methods of production. However, the life cycle assessment for crop-derived hydrogen also carries the entire impact of the crop itself, which ultimately depends on several other factors such as type of crop, techniques of irrigation, local production, and other geographical aspects [4]. In addition, an environmental assessment can be carried out through different perspectives and using different footprints, which, as will be further discussed, can result in different results depending on the type of impact that is being evaluated [12]. The renewable aspect of crop-derived hydrogen may decrease its carbon footprint when the entire life cycle is considered, but agricultural products also require high use of land and water that is usually not required for extracting fossil fuels, which may counterbalance the environmental benefits of these alternative pathways.

In this sense, this chapter provides a summarized discussion about the current hydrogen industry, mentioning its strong reliance on fossil-based sources that may end up decreasing the alleged environmental benefits of using this supposedly carbon-free fuel. Both traditional and alternative production pathways are presented and discussed, alongside the data that is currently available regarding their environmental impact through the use of different environmental footprint indicators. Finally, the main barriers to be overcome in order to move toward a more renewable hydrogen matrix are discussed, including possible technologies that could be used in this transition.

2 Overview of the Current Hydrogen Industry

Hydrogen use as a fuel has been reported with great potential and perspective for the future years. Some numbers support the emerging character of this market: first, the world consumption, which has grown 11.5 Mt from 2010 to 2018 [15], and is expected to present a 16.1 Mt increase until 2030 [16]; besides that, it is estimated that the hydrogen market mobilized around 160 billion dollars yearly worldwide [17].

Although displaying sustainability banners, and recurrently being associated with energy transition and circular economy, the current market is more gray than green. Such expression is supported by the color scale that indicates which is the route used

Fig. 1 Hydrogen production route color scale. *Source* authors

for hydrogen manufacture. Figure 1 presents the color scale and its correspondent production pathways.

The current scenario for the hydrogen industry relies mainly on methane steam reforming, a traditional process that accounts for 73% of all production units [6]. Such widespread technology is due to extended know-how and infrastructure available since the 1960s, when steam reforming replaced water electrolysis as the least expensive way to produce hydrogen, with the possibility of synthesizing hydrogen with an energy efficiency of 76% [18]. The current problem with the hydrogen scenario is in the renewable character promised and not delivered, as only 2% of the world's hydrogen supply comes from non-fossil feedstock [15]. This reflects in 830 Mt of CO_2 emissions being released yearly, almost twice the amount emitted in the entire Brazilian territory [19].

Around the uncertainties still around water electrolysis, such as technology development, social acceptance, costs, and international policies toward climate and energy, green hydrogen is a carrier that remains in stand-by on industrial and transportation sectors [20]. A mid-term solution to supply demand and decrease CO_2 emissions would be steam reforming allied with carbon capture technologies [21], an approach labeled as the best trade-off for industry between technical, economic, and environmental output by Al-Qahtani et al. [18]. Biogenic feedstock, such as crops and waste for hydrogen production, can result in a wide range of emissions. The use of energy crops is intermediary between gray and green hydrogen, concerning greenhouse gas emissions; however, the use of wastes can yield negative emissions, when addressed along with carbon capture [22]. Nevertheless, the use of energy crops has its limitations, and to apply waste material there is a need to deploy an effective supply chain network.

Mid-term action toward decreasing carbon footprint for hydrogen must rely on carbon capture, utilization, and storage technologies (CCUS) retrofitted to steam methane reforming (SMR). Espegren et al. observed that the blue hydrogen production should be viable where the natural gas prices are lower, such as in the Middle East, and the economic performance of such plants would vary with this utility cost around the globe [20]. Addressing cost reduction in the production chain of blue hydrogen, Ali Khan et al. established that the captured CO_2 should go through further upgrading into high-value products [23]. This action would help generate more income and fewer costs with geological storage.

Moreira dos Santos et al. proposed a strategy to improve blue hydrogen economic performance by valorizing the methane industry, taking advantage of the existing infrastructure to comply with current and future energy demands [24]. Since IEA reports that blue and gray hydrogen do not have much different production costs [15], any strategy aiming to decrease blue hydrogen costs, in the eyes of industry, is essential for the transition that would eventually culminate on green hydrogen being feasible. Similarly, the economic viability of hydrogen produced from crop-derived feedstock could be increased through the use of more interconnected cycles, as shown in a study of a cogeneration system for glycerol-derived hydrogen coupled with a soybean biodiesel facility [11].

3   Hydrogen from Crops: Thermochemical Processes

As the hydrogen industry is currently dominated by the steam reforming of natural gas [5], the benefits of using this carbon-free fuel are decreased when the entire life cycle is taken into account [2]. Steam reforming is the most traditional of the many existing thermochemical processes for hydrogen production, which also includes variations such as dry reforming, autothermal reforming, mixed reforming, gasification, and pyrolysis. All of these processes are considered "thermochemical" because they promote the breaking of feedstock's molecules under high temperature in order to release the molecular hydrogen present in the fuel that is being consumed.

Despite the dominance of fossil-based feedstock in the majority of hydrogen-producing facilities, the thermochemical pathways present the flexibility of using the same technology to convert biomass-derived materials, offering a possible solution to some of the environmental footprints associated with hydrogen production from natural gas. Following, some of the main thermochemical processes are discussed, including both the current, fossil-based, and alternative, crop-derived pathways.

3.1   *Steam Reforming*

Steam reforming is the most consolidated hydrogen-producing pathway, representing the vast majority of this fuel's production. As with every thermochemical process, steam reforming promotes the decomposition of molecules under high temperature and releases hydrogen, with the particularity of being carried out in the presence of steam. This process is mostly endothermic, i.e. it consumes heat to carry out the reactions, and therefore an external source of energy is required [11]. Natural gas is the most common feedstock used in this process due to its high availability and low costs. This process also includes a secondary reactor that boosts hydrogen production even further through a reaction called water gas shift, which consumes water to convert carbon monoxide into carbon dioxide and hydrogen.

Besides the traditional use of fossil fuels, this process can also be used to decompose other feedstocks. A large variety of crop-derived substances such as glycerol, ethanol, butanol, and vegetable oils have already been successfully used for hydrogen production through steam reforming. The major advantage of these alternative pathways is the adaptability to existing infrastructure since the process can run with different feedstock with minor modifications. Common changes between steam reformers consuming traditional and alternative fuels include operational conditions, such as temperature and steam/fuel molar ratio, and type of catalyst.

The system's performance also varies according to the feedstock that is used, which affects the theoretical maximum hydrogen production per mole of fuel and the optimal conditions of operation. Table 1 presents the stoichiometric steam reforming reaction, as well as common values for temperature and steam/feedstock ratio for some of the main fuels used for hydrogen production through this process.

Table 1 Simplified steam reforming reactions and common operational points

Feedstock	Simplified reforming reaction	T (°C)	S/F (−)
Natural gas/methane [25]	$CH_4 + H_2O \leftrightarrows$ $CO + 3H_2; \Delta H°_{298} = 251 \frac{kJ}{mol}$	700	3.2
Glycerol [26]	$C_3H_8O_3 + 3H_2O \leftrightarrows$ $3CO_2 + 7H_2; \Delta H°_{298} = 128 \frac{kJ}{mol}$	723	6.0
Ethanol [27, 28]	$C_5H_6O + 3H_2O \leftrightarrows 2CO_2 + 6H_2$	800	4.0
Butanol [29]	$C_4H_{10}O + 7H_2O \rightarrow 4CO_2 + 12H_2$	500	2.8

Among the crop-derived substances used for hydrogen production, one of the most promising options is glycerol. Glycerol is a largely available by-product of the biodiesel industry [30], therefore being indirectly produced from vegetable oils (and therefore from crops). The increasing quantity of this product makes it relatively cheap, despite its reforming process being associated with higher hydrogen production costs when compared to the traditional steam methane reforming [11]. This process can also be carried out in co-generation systems, co-producing heat, power, and hydrogen [11, 31]. Finally, as steam reforming can use crude glycerol directly from biodiesel facilities, it does not require further purification costs for valorization of this by-product, ultimately decreasing its final hydrogen production costs [2].

Another important feedstock for biomass-based hydrogen production from steam reforming is ethanol. This fuel, mostly derived from crops such as sugarcane and corn, is largely available around the world, playing an important role in the transportation matrix of several countries. Since the ethanol industry already presents a strong logistic for producing and distributing fuels, using this crop-derived alcohol for hydrogen production can be an option for quickly diversifying the fuel matrix [4]. Moreover, as second-generation ethanol can be produced from lignocellulosic material, the sugar and alcohol industry can also produce hydrogen without a direct competition for food and biofuels. Steam reforming is and should continue to be the most common method for hydrogen production. Its feedstock flexibility allows the use of a very high variety of fuels, including many crop-derived options. Besides the highlighted use of glycerol and ethanol due to its direct link with the fuel industry, successful hydrogen production through steam reforming was already reported for methanol [32], butanol [33], biomass tar [34], bio-oils [35], and many more options and improvements should keep being presented in the next years.

3.2 Dry Reforming

As it happens in the steam reforming process, dry reforming also decomposes molecules to release hydrogen gas. However, opposite to the traditional process, dry reforming does not require steam and can operate at atmospheric pressure, which simplifies the construction of the reforming system [36]. In addition, this process can convert some CO_2 into syngas but present higher coke formation when compared to other reforming processes [37]. Dry reforming also offers high feedstock flexibility, with reported uses of several alkanes (methane, ethane, butane), alcohols (ethanol, methanol, glycerol), and other hydrocarbons [38], many of which can be produced from crops.

3.3   Partial Oxidation and Autothermal Reforming

Despite being the most common way of producing hydrogen because of its high hydrogen yield, steam reforming is strongly endothermic and presents a high consumption of heat and steam. As an option to produce hydrogen while avoiding this problem, a process called partial oxidation can be carried out [39]. This process oxidizes part of the feedstock, releasing heat and achieving high temperatures without the need of an external source of energy [40]. Despite being more simple and highly exothermic, partial oxidation presents the lowest hydrogen yield among the common reforming processes, therefore being limited to specific applications [39].

Processes that involve partial oxidation, such as gasification of biomass, can produce hydrogen from almost any crop-derived feedstock. Due to its flexibility, this process is commonly applied to non-edible residues, such as straws [41], husks [14], and leaves [42], as well as wood and other crops that are not directly linked to the food industry [43]. In addition to the use of basic feedstock, gasification is also an option for producing hydrogen from more processed crop-derived products, such as phenols and alcohols that may originate from cellulose and hemicellulose materials in several industrial sectors [44].

As an intermediate option between the efficient but energy-consuming steam reforming and the inefficient but exothermic partial oxidation, a process called autothermal reforming can be applied. This pathway combines the endothermic steam reforming and the exothermic partial oxidation reactions, reaching a nearly neutral heat balance [39]. In other words, the oxidized fraction provides the heat that ensures that the steam reforming takes place, producing less hydrogen than in pure steam reforming but without the need of consuming heat from external sources [45]. As occurs with partial oxidation, autothermal reforming can also use a wide variety of crops and their by-products, being therefore a very adaptable process for producing hydrogen from agricultural residues [46].

3.4   Pyrolysis

Pyrolysis is a process that consists of the thermochemical decomposition of a substance in the absence of oxygen, producing solid, liquid, and gaseous products [47]. The selection of feedstock, particle size, carrier gas flow, and temperature strongly affects the process outputs, in such a way that proper control of the operational parameters can be used to maximize hydrogen production [48]. This process presents some variations such as slow, fast, vacuum, and flash pyrolysis, which varies according to its heating rate, average temperature, residence time, and desired products [49].

In addition to the hydrogen that is directly produced in this thermochemical process, pyrolysis can also yield valuable liquid fuels and char that can be either used for specific applications [50] or upgraded to hydrogen through reforming of pyrolysis oils, sometimes even using the resulting biochar as a catalyst support [51]. In this sense, pyrolysis is extremely adaptable for producing hydrogen and other biofuels from crops, being capable of using virtually any biomass-derived material, ranging from wood [52] to leaves [53], seeds [54], and bio-oils [35].

4 Hydrogen from Crops: Biological Processes

Biological processes for hydrogen production represent a more sustainable alternative for H_2 production when compared to currently available thermochemical approaches, with advantages such as low energy consumption to impose the conditions for the progress of the process, the possibility of direct use of renewable raw materials as a substrate [55, 56], and can also be an alternative to simultaneously address the need to increase energy supply and improve waste management through a waste-to-energy (WTE) supply chain [57].

The biological processes for obtaining hydrogen are carried out by different anaerobic microorganisms, such as bacteria and algae. The process can be classified as biophotolysis of water that uses algae; photodecomposition of organic compounds by photosynthetic bacteria; dark fermentation; or hybrid systems [58]. Phototrophic biological pathways for hydrogen production have important operational difficulties such as the complexity of controlling penetration and uniform distribution of light in the reactor. In addition, they tend not to be cost-effective due to the need for artificial light. In this sense, dark fermentation is currently the biological process most studied and with the highest potential for obtaining hydrogen on a larger scale [59]. The hydrogen production via dark fermentation is considered the most promising sustainable route to obtain this resource; its net energy ratio, which consists of the ratio between hydrogen output (MJ) and non-renewable energy input (MJ), has a value of 1.9, while for steam methane reforming this ratio is equal to 0.64, where values greater than 1 indicate the renewable nature of the process [60].

Biohydrogen production via dark fermentation occurs through metabolism mainly of strictly anaerobic and/or facultative anaerobic bacteria; pure culture of bacteria (e.g. *Enterobacter* sp., *Clostridium* sp.) in general results in higher hydrogen yield, but utilization of mixed culture offers the possibility of using more complex and varied substrates and is also easier to control [61, 62]. Acidogenic fermentation of carbohydrate-rich materials results in gaseous products such as hydrogen, and liquid effluent containing organic acids (acetic acid, butyric acid, propionic acid, among others) as well as solvents such as alcohols and ketones [10]. The formation of hydrogen results from metabolic biochemical reactions of different species of bacteria whose prevalence depends on the operating conditions of the process. In general, the process is summarized in the oxidation of organic substrates releasing electrons, which needs to be disposed to maintain electrical neutrality, and in the absence of external acceptors of electrons, protons to act as electron acceptors being reduced to molecular hydrogen (H_2) [58, 63].

Acidogenic fermentation directly uses simple sugars such as glucose, however, not all carbohydrates present in substrates are in the form of monomers, so the hydrolysis step that involves breaking chemical bonds of complex insoluble compounds leading to simpler soluble compounds is fundamental, and in some cases, this can be a limiting step in the process due to the low speeds at which it occurs [64].

The substrate metabolization for hydrogen generation occurs initially via glycolysis, converting glucose into pyruvate. From pyruvate, there is the formation of acetyl-CoA, which is a fundamental intermediate in the hydrogen production process. The formation of acetyl-CoA can be catalyzed by the enzyme pyruvate-ferredoxin oxidoreductase (PFOR), as per Eq. 1, which requires the reduction of ferredoxin (Fd). Fd also is reduced in the reaction with NADH conducted by NADH-ferredoxin oxidoreductase (NFOR), and by the action of hydrogenases there is the process of reducing H^+ protons using Fd electrons, which results in the formation of molecular hydrogen (Eq. 2) [62, 65]. The acetyl-CoA can then be metabolized to acetate or butyrate, and in the case of acetate formation an additional mole of H_2 is produced per mole of NADH generated in glycolysis being reduced to NAD^+. This fact does not occur in the butyrate formation pathway, in which NADH is reduced to NAD^+ by the oxidation of acetoacetyl-CoA to butyrate. The net balance of hydrogen production is therefore 4 mol H_2/mol glucose when acetate is the final product (Eq. 3) and 2 mol H_2/mol glucose for butyrate (Eq. 4) [58, 66].

$$\text{pyruvate} + \text{CoA} + \text{Fd}_{ox} \leftrightarrow \text{acetylCoA} + \text{Fd}_{red}\,\Delta G^\circ = -19.2\,\text{kJ mol}^{-1} \quad (1)$$

$$2H^+ + \text{Fd}_{red} \rightarrow H_2 + \text{Fd}_{ox} \quad (2)$$

$$C_6H_{12}O_6 + 2H_2O \rightarrow 2CH_3COOH + 2CO_2 + 4H_2 \quad (3)$$

$$C_6H_{12}O_6 + 2H_2O \rightarrow CH_3CH_2CH_2COOH + 2CO_2 + 2H_2 \quad (4)$$

The formation of acetyl-CoA can also be driven by the action of the enzyme pyruvate-formate lyase (PFL), with the joint formation of formate, as in Eq. 5. In this pathway, hydrogen is released along with CO_2 when formic acid is further broken down by the activity of the enzyme formate-hydrogen-lyase (FHL) (Eq. 6) [67].

$$\text{pyruvate} + \text{CoA} \leftrightarrow \text{acetylCoA} + \text{formate}\,\Delta G^\circ = -16.3\,\text{kJ mol}^{-1} \quad (5)$$

$$\text{formate}(\text{HCOO}^-) \rightarrow H_2 + CO_2 \tag{6}$$

In practice, the formation of a mixture of different metabolites not accompanied by the generation of hydrogen at the end of the process leads to a reduction in the yield of H_2 production to a range of 1–2.5 mol H_2/mole glucose [65], which leads to the identification of by-products such as propionic acid, lactic acid, and solvents, as unwanted, taking into account the focus on obtaining a higher yield of H_2 and reaching values close to the so-called Thauer limit [68], which predicts a maximum theoretical production of 4 mol H_2/mole glucose, obtained in the aforementioned pathway in which acetate is the by-product formed (Eq. 3) [62, 69]. When xylose is taken as a reference as the primary substrate of the process, similar to the case of glucose, there is a theoretical limit of 3.33 mol H_2/mole xylose when acetate is the fermentative product and 1.66 $_{mol}$ H_2/mole xylose in the case of the butyrate [55].

The raw materials cost plays a fundamental role in the economic feasibility of these hydrogen production processes, and in this sense, several waste materials have been successfully used in different hydrogen production processes, such as in the biological approach [58]. Lignocellulosic biomass has a high potential for use in hydrogen production via fermentation, as they are carbohydrate-rich, abundant, potentially biodegradable, and cheap substrates [70], which in this group include agricultural residues, municipal solid wastes, industrial wastes, food wastes, and others [10, 71].

Agricultural residues from harvested crops are the substrates with the highest potential for use in obtaining renewable energy due to their abundant availability, with an estimated global annual yield higher than 200 billion dry tons [72]. Agricultural crops such as cereals, legumes, oilseeds, sugarcane are commodities that have the most produced agro-industrial wastes in the last few years [73]. The residues from harvested crops include pells, husks, seeds, cobs, pulps, press cakes, leaves, stalks, bagasse, among others [10], and such raw materials are constituted of a complex lignocellulosic matrix composed of hemicellulose, cellulose, and lignin. The arrangement of the molecules of these components and the connection between them creates a rigid three-dimensional matrix, with a characteristic of difficult degradation [74, 75]. Different types of agricultural wastes show variation in the percentage distribution of their composition. The literature reports that cellulose is the main component ranging from 35 to 50%, hemicellulose 20 to 35%, and lignin 10 to 25% [72].

The recalcitrant characteristic of these raw materials due to their structure with a lignocellulosic matrix is the main bottleneck for hydrogen production since the hydrolysis to release fermentable sugars becomes the limiting step in the process, substantially reducing the process yield in relation to the theoretical limit [76]. In an attempt to improve the low hydrogen production from lignocellulosic biomass, a raw material pretreatment step has been proposed as an alternative. The applied process aims to break the recalcitrant structure and release the carbohydrate molecules in the solution and, in addition, break down the crystalline arrangement of the cellulose molecule and depolymerization, making the substrate accessible for enzymatic hydrolysis and making it assimilable to fermentative microorganisms [77, 78].

Several pretreatments have been tested, such as physical, chemical, physicochemical, and biological, and studies have shown that the most convenient one varies according to the composition of the raw material [79]. The choice of a pretreatment method for a specific lignocellulosic waste must consider important aspects such as minimizing carbohydrate degradation and avoiding the formation of inhibitory compounds such as furfurals that can harm the process. Furthermore, the pretreatment process is expensive and can reduce the economic feasibility of implementing a biological hydrogen production system, in this sense, it would be beneficial to avoid pretreatment [55, 62].

Table 2 shows some recent studies conducted to assess the potential and find better conditions for hydrogen production via the biological process from various agricultural wastes and contribute to the advancement of knowledge about strategies to maximize the energy recovery through the biological production of hydrogen.

The main challenges regarding the biological production of hydrogen via fermentation are related to the theoretical limit for the hydrogen yield, which represents a barrier to the implementation of this technology on an industrial scale, since hydrogen yields and conversion yield from lignocellulosic biomass reach values well below the theoretical limit of 4 mol H_2/mol glucose, due to the complex and polymeric composition of the substrate. Also due to mixed cultures as inoculum that provide for the development of several metabolic routes different from the acetate production

Table 2 Biohydrogen production from different agricultural residues

Substrate agricultural residues	System mode	Substrate pretreatment	H_2 yield	References
Rice straw	Anaerobic baffled reactor	Alkaline pretreatment with ammonium hydroxide	1.19 mol H2 mol glucose^{-1}	[80]
Sugarcane bagasse	Batch reactors	Non-applied	1.96 mol H2 mol sugar^{-1}	[81]
Corn stalk pith	Batch reactors	Hydrolyzed with commercial cellulase at 50° C	2.6 mol H_2 mol sugar^{-1}	[82]
Corn stover	Batch reactors	Steam explosion (90–220 °C, 3–5 min)	3.0 mol H_2 mol hexose^{-1}	[83]
		Acid-thermal hydrolysis H_2SO_4 0.25–4(v/v), 121 °C, 30–180 min	2.24 mol H_2 mol hexose^{-1}	[84]
Cellulose	Batch reactors	Non-applied	2.71 mol H_2 mol hexose^{-1}	[85]
Solid carob waste	Batch reactors	Particle size reduction	1.37 mol H_2 mol hexose^{-1}	[86]

pathway, in this sense large-scale processes of this type require very large volume reactors [10, 70].

5 Electrolysis as an Alternative for Green Hydrogen Production

Water electrolysis is an alternative that does not associate with the main environmental impacts related to energy crops for energy, such as direct and indirect use of land or carbon footprint. This is due to electrolysis character, an electrochemical reaction that splits the water molecule into oxygen and hydrogen gas via electricity.

The water electrolysis technology is promising because it converts renewable energy into a chemical energy carrier, which is easy to store, facilitates control of the grid, and additionally produces high purity hydrogen [87]. In addition, water electrolysis can help bring renewables to sectors that electrification cannot reach, such as chemical synthesis that involves H_2 [15], the transportation sector (especially heavy-duty and long-distance) [88], and eventually the steel segment, by direct reduced iron (DRI) method that can potentially decrease 740 Mt of CO_2 emissions yearly [15].

The efficiency of electrolyzer systems today ranges between 60 and 81% depending on the technology type and load factor [89]. Producing all of today's dedicated hydrogen output (69 Mt) from electricity would result in a demand of 3600 TWh. Moreover, electrolysis requires water as well as electricity, and around 9 L of water are required to produce 1 kg of H_2. Therefore, if a full 70 Mt of hydrogen production were to be produced by electrolysis, approximately twice the current water consumption for hydrogen spent on steam methane reforming would be used [15].

There has been an increase in new electrolysis installations over the last decade, aiming to produce hydrogen mostly in Europe, although projects have also started or were announced in Australia, China, and the Americas. Three electrolyzer technologies are available currently: alkaline electrolysis, proton exchange membrane (PEM) electrolysis, and solid oxide electrolysis cells (SOECs). The alkaline electrolyzers are fully mature as a technology and have a longer lifetime. The PEM technology was introduced in the 1960s by General Electrics, was quickly developed, and most recently reached full development [90]. SOEC technology is a few stages below in technology development. This electrolyzer demands high thermal and electrical energy, besides high operating temperature. Consequently, the total energy is constant, and with increasing temperature, required electricity reduces and required thermal energy increases [91]. Moreover, the highlighted challenge relies on selecting materials to resist high temperatures with competitive costs (IEA, 2019).

In this sense, alkaline electrolyzers are most present in the industry today. A survey of power-to-gas plants in Europe showed that 21 plants operate with alkaline electrolyzers, followed by 12 PEM, 4 SOEC, and 3 combining PEM and alkaline technologies [92]. An overview of the main technical and economic characteristics of each electrolysis technology is presented in Table 3.

Just expanding water electrolysis use into the market would be ineffective on decarbonization terms, which is due to the carbon footprint designated by the current global electric matrix. The global warming potential (GWP) category of life cycle impacts for electrolysis using power from the grid (11.1 kg CO_2e/kg H_2) is comparable to the numbers of steam methane reforming without CCUS (9.0–11 $_{kg}$ CO_2e/kg H_2). Nevertheless, if renewable energy is used, the carbon footprint decreases by 90 and 95% for solar and wind power systems, respectively [22].

Hence, water electrolysis challenges until mid-century passes through making world electricity greener. The reduction of renewable electricity costs and in consumption of water for electrolysis is mandatory for scale-up operations, as much as finding ideal materials for the production of electrolytic cells, and also establishing a large-scale electrolysis supply chain system [94].

6 Environmental Footprints of Hydrogen Production

Understanding the environmental footprints of hydrogen production is one of the key challenges to decrease the environmental impacts of this highly fossil dependent, yet supposedly green fuel. Both technological and economical aspects favor the traditional methane (natural gas) steam reforming, but when environmental factors are taken into account, alternative production processes may become more desirable [2].

This item discusses the main pathways for hydrogen production, i.e. methane steam reforming, biomass steam reforming, green, blue and biological hydrogen, using various footprint indicators, such as global warming potential, water footprint, land use, and more generic indicators. Since environmental impacts strongly depend on the process, type of feedstock used, and geographical aspects, each one of the main processes for hydrogen production will be separately discussed, including different environmental footprints that may ultimately offer a complete overview of the impact behind the hydrogen industry.

Life cycle assessment (LCA) has been widely used to evaluate the environmental impact analysis of hydrogen production technologies. This methodology consists of aggregate inventory data, and most often used system boundaries, which is cradle to gate, i.e. the resource extraction, production and supply of hydrogen to the end-users, and convert the results into impact categories. Life cycle impact assessment tools were developed to help elaborate the environmental assessment of a product by calculating indicator scores for processes or products [95]. The Eco-Indicator 99 (Eco-99) is an example, which helps identify environmental damage in human health, ecosystem quality and resources; as there is the successor ReCiPe method, which provides scores at two levels midpoint (18 categories) and endpoint (3 categories) [96], similar to IMPACT 2002+ that provide results in four endpoint categories and 15 midpoint categories [97].

Table 3 Techno-economic review of electrolyzers [15, 89, 93]

	Alkaline electrolyzer			PEM electrolyzer			SOEC electrolyzer		
	Present	2030	2050	Present	2030	2050	Present	2030	2050
Technology Readiness Level	9 (Mature)			9 (Commercial)			6 (Demonstration)		
Electrolyte	KOH (Liquid)			Polymer (solid)			Ceramic (solid)		
Charge carrier	OH^-			H^+			O^{2-}		
Operating pressure (bar)	1–30			30–80			1		
Operating temperature (°C)	60–80			50–80			650–1000		
Conversion efficiency (kWh/kg H_2)	51			58					
Density (A/cm²)	0.2–0.4			0.6–2			0.3–1		
Power Consumption (kWh/Nm³)	4.5–7			4.5–7.5			>3.7		
Electrical efficiency (% LHV)	63–70	65–71	70–80	56–60	63–68	67–74	74–81	77–84	77–90
Range of operation (% of nominal load)	10–110			0–160			20–100		

(continued)

Table 3 (continued)

	Alkaline electrolyzer			PEM electrolyzer			SOEC electrolyzer		
	Present	2030	2050	Present	2030	2050	Present	2030	2050
Hydrogen production capacity (Nm³/h)	<7–60			<40			<40		
CAPEX (USD/kW)	500–1400	400–850	200–700	1100–1800	650–1500	200–900	2800–5600	800–2800	500–1000
OPEX (% of initial CAPEX/year)	2%			2%			2–4%		
System lifetime (years)	20			20			3–6	12–20	20
Advantages	o Low capital cost o High stability o Mature technology o Longer lifetime o Lower cost catalyst			o Compact design o High density o High purity o Quick start time o Simple design			o Lower energy need o High current density o High efficiency o Lower cost catalyst		
Disadvantages	o Corrosive character of electrolyte o Lower density; slow			o High cost for membrane o Lower durability o Costly catalyst			o Instable electrode o Safety issues o Sealing difficulties o Delamination of electrodes		

6.1 Gray, Blue, and Green Hydrogen

Considering the assessment of the environmental impacts of hydrogen-producing pathways, recent studies have been dedicated to evaluate such footprint [18, 98–100]. Gray hydrogen provides the worst environmental impacts on ozone depletion, particulate matter, and photochemical ozone formation. However, it is the closest method to be projected as an ideal situation when addressing categories such as ionizing radiation, terrestrial, and freshwater eutrophications [93]. However, these are not the main categories that hydrogen production has been evaluated. Typically, resource consumptions, energy requirements, and emissions are the life cycle point of views investigated, due to the importance of climate change mitigation policies in decision-making, and for climate change, the gray hydrogen causes the highest impact, as stated in [95].

Through Eco-99 environmental indicator, the environmental impact associated with gray hydrogen production was reported around 951 mPt/kg H_2 [25]. For comparison purposes, the use of glycerol instead of natural gas for hydrogen production through steam reforming can reduce this value up to 282 mPt/kg H_2 [2], which highlights a potential impact reduction of 70% to be achieved in the hydrogen industry. However, considering only the impact associated with water consumption through the water footprint indicator, gray hydrogen actually presented a lower impact even when compared to biomass-based pathways, with only 0.257 m^3/kg H_2 against 0.77 and 9.65 m^3/kg H_2 estimated for glycerol and bioethanol steam reforming, respectively [4].

The blue hydrogen production chain can be controversial when addressing its environmental load, especially GWP. The differentiation from gray hydrogen is in the carbon dioxide captured from SMR process, and sometimes also from the flue gases created by burning the natural gas that provides heat and high pressure, which drives the SMR process. However, a third source of CO_2 that is typically not captured is the carbon footprint from the grid electricity, which is a current problem concerning the global warming potential of blue hydrogen. In this sense, the work of Howarth and Jacobson [101], despite stating that there is an importance in natural gas and blue hydrogen for the next 20 years accommodating energy transition, shows that the numbers brought in their study reinforce that in a low-carbon economy no room would be available to blue hydrogen. This is justified once that GWP impacts indicated the need for renewable electricity in the steam reforming process in order to achieve effective lower emissions, which, however, did not make sense once these resources would be better used to produce green hydrogen.

Moreover, methane steam reforming presented lower water consumption potential when compared to coal gasification and several electrolysis and biological production processes, while presenting slightly worse indicators when compared to certain pathways of biomass gasification and reformation [12]. Water consumption and water-related effects are emerging categories with the greatest interest to the life cycle analysis (LCA) that includes water-intensive hydrogen production chains like electrolysis. And in this sense, Mehmeti et al. [12] have shown that the most modern

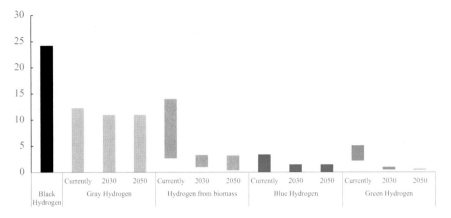

Fig. 2 Reported values for GWP of several H_2 producing pathways (kg CO_2eq/kg H_2) [12, 22]

electrolyzers such as SOEC can provide lower water footprints than the traditional PEMs.

When addressing hydrogen production via water electrolysis and renewable sources of energy, the greatest impact in terms of GWP is due to the manufacturing of the power systems (wind turbines and solar panels). Nevertheless, the hydrogen synthesis from water electrolysis delivers 970 g CO_{2e}/kg H_2 and 2412 g CO_{2e}/kg H_2 for solar and wind electricity, respectively, values that are 92 and 79.7% smaller than the steam methane reforming [102]. Hydrogen production through electrolysis should be able to remove up to 938 Gt of CO_2 yearly worldwide as estimated by [103], as suggested in Fig. 2.

However, the trade-off is with the economic viability, which due to hydrogen production capacity is biased to SMR that can produce almost a thousand times more hydrogen in a day than water electrolysis [102]. Future directions seem to rely on improvements in renewable hydrogen and energy technical efficiency and a subsequent decrease in feedstock/energy and labor requirements. In this sense, the renewable hydrogen options might potentially achieve a fully sustainable performance [104]. However, policymakers action should be aware of this occurrence, because right now the ambition of the Paris Agreement seems to be far from being accomplished, and hydrogen, which is supposed to play a key role in decarbonization, does not have a well-established carbon-limited production.

6.2 Hydrogen from Biomass

Due to the renewable aspect of biomass, it is natural to expect that using this feedstock for producing hydrogen can potentially reduce the environmental footprint of the hydrogen industry. In fact, several hydrogen production pathways using biomass presented better environmental indicators when compared to the traditional methane steam reforming, such as lower global warming potential, freshwater eutrophication potential, water scarcity potential, fossil scarcity potential [12], and performed better when using more generic indicators such as Eco-95 [25] and Eco-99 [2]. When a single and more generic indicator is used, the fact that natural gas is a nonrenewable fuel with high global warming potential weights is in favor of biomass-based hydrogen, but this is not true when more specific indicators are used. Depending on the type of process and biomass used, producing hydrogen from crops may represent a burden on hydric resources, even when the production pathway is set to be as renewable as possible. Actually, a recent study reported that producing hydrogen from bioethanol steam reforming using sugarcane bagasse as a heat source could consume over 37 times more water than traditional processes, which is due to the high amounts of water demanded for irrigation [4]. The same applies to land use, which tends to be significantly higher for biomass-based hydrogen due to the area used for feedstock production [12]. In brief, the use of biomass for producing hydrogen can significantly decrease the environmental impact of this fuel from a global warming point of view, but also demands an intensive use of resources such as water and land, as the Hydrogen Council reported that accounting for gross water demand the use of energy crops for steam reforming is the worst scenario, along with biomass gasification [22]. Therefore, the environmental benefits should be properly weighted before choosing the best options for a cleaner hydrogen matrix.

6.3 Biological Hydrogen

Most of the studies of hydrogen production from biological pathways focus on variations of dark fermentation. In a study comparing environmental footprints of different dark fermentation processes, Mehmeti et al. [12] reported that this option presents a much higher water consumption when compared to most of the existing pathways (except for some electrolysis technologies), while also presenting a land footprint higher than the traditional methane reforming but lower than certain biomass-based pathways. The same study indicated that, among the biological processes included in the analysis, the dark fermentation with microbial electrolysis cell and energy recovery presented the lowest water and land footprints per kg of H_2, and was the only of these processes that resulted in a global warming potential lower than conventional pathways. Another efficient way to produce hydrogen using a combination of biological and thermochemical processes was environmentally assessed by Battista et al. [105]. The authors compared three pathways to produce hydrogen from biogas, including autothermal reforming, steam reforming, and an ICE/electrolyzer combination, concluding that the first two options were comparable in terms of global warming potential and global energy requirement, while the latter presented worse indicators for almost all footprints included in the assessment.

7 Conclusions and Key Challenges for a Greener Hydrogen Matrix

The hydrogen market has been transitioning, and the problems concerning the emissions designated by steam methane reforming need a solution. Therefore, alternatives, strategies, and policies must be studied and applied toward a greener matrix. The main transition driver to industry seems to be blue hydrogen; nevertheless, technologies to capture CO_2, despite being promising, demand high capital expenditure and involve high operation costs [106]. Moreover, the current grid situation would provide electricity into the blue hydrogen production chain that would address undesirable environmental loads, which would make its emissions closer to gray hydrogen [101].

It is common sense that the market for green hydrogen has gained attention through sustainable policies looking up to the economic recovery in a post-pandemic scenario [107]. The synthesis of blue hydrogen might be placed only as a tool for transition because green hydrogen production is associated with more opportunities and perspectives for the future. Therefore, the valorization of different production chains is an economic driver for the hydrogen market, together with public policy. The use of crop-derived hydrogen has the same drawbacks as first-generation biofuels addressed, land-use, water consumption, and food-energy conflict are coherent arguments to bypass such technology in order to guarantee food and energy security, besides environment preservation. In this sense, the alternative technologies for H_2 production must be approached in terms of identifying its advantages.

In scenarios of accelerated development and cost decline, investments in the combination of renewable and intermittent energy and water electrolysis might emerge as key for green hydrogen. These configurations are called Power-to-Gas and Power-to-X, which have been widespread in Europe, mainly in Germany, where the expectations are to have an installed capacity between 6 and 16 GW_{el} by 2050 [108]. Being, in consequence, better alternatives than cultivation, due to life cycle impacts that are designated by the practice. It was reported by Ozturk and Dincer [93] that Power-to-Gas is the option where 4 out of the 8 environmental impacts investigated were closer to the ideal situation.

The power-to-gas concept uses renewable or surplus electricity for hydrogen production through water electrolysis. Additionally, the use of such H_2 to produce different products is called Power-to-X. The main goal is to store energy from renewable sources by converting it into carriers of easy storage, which would be important to balance the grid operation and to take renewables into sectors of industry that are difficult to electrify [92]. Besides that, energy storage is expected to play a crucial role in the development of a low-carbon energy model in EU countries [109]. A typical Power-to-X system is the production of hydrogen and its upgrading to methane, or Synthetic Natural Gas (SNG) by Sabatier reaction [110]. By this, more renewable energy is inserted into the gas network, making it possible to explore the transportation infrastructure via pipeline network, once that for hydrogen only fractions smaller than 10% in volume are possible [111]. Moreover, starting in H_2 to CH_4 conversion, the discussion opens to synthesize different products, such as hydrocarbon-based renewable fuels, chemicals, heat, and power. This would highlight the capacity of integrating key sectors for the decarbonization of the economy, such as gas, power, and transport.

The future of hydrogen relies on governmental strategies to support the action of the economic drivers, which comprises the use of crops. An example is the National Plan for Hydrogen Consolidation in Brazil [112], which suggests inserting ethanol, a highly available hydrocarbon produced from sugarcane and sugarcane-based second-generation biomass, in the hydrogen production chain as a feedstock for steam reforming and direct oxidation in fuel cells.Other works made the characterization of hydrogen production processes and suggestions for the incorporation of steam reforming of ethanol for the hydrogen production process in sugarcane industry, as regards to Brazilian conditions [113–115]. Additionally, the use of hydrogen as the main carrier and the fit with renewables intermittence has yielded it the label of "the new oil". And, by this, the economy may rely more and more on the hydrogen industry in the next decades.

References

1. Rievaj V, Gana J, Synák F (2019) Is hydrogen the fuel of the future? Transp Res Procedia 40:469–474. https://doi.org/10.1016/j.trpro.2019.07.068
2. Rocha DHD, de Souza TAZ, Coronado CJR, Silveira JL, Silva RJ (2021) Exergoenvironmental analysis of hydrogen production through glycerol steam reforming. Int J Hydrog Energy 46:1385–1402. https://doi.org/10.1016/j.ijhydene.2020.10.012
3. Guban D, Muritala IK, Roeb M, Sattler C (2020) Assessment of sustainable high temperature hydrogen production technologies. Int J Hydrog Energy 45:26156–26165. https://doi.org/10.1016/j.ijhydene.2019.08.145
4. de Souza TAZ, Rocha DHD, Julio AAV, Coronado CJR, Silveira JL, Silva RJ et al (2021) Exergoenvironmental assessment of hydrogen water footprint via steam reforming in Brazil. J Clean Prod 311. https://doi.org/10.1016/J.JCLEPRO.2021.127577
5. Pashchenko D (2019) Experimental investigation of reforming and flow characteristics of a steam methane reformer filled with nickel catalyst of various shapes. Energy Convers Manag 185:465–472. https://doi.org/10.1016/j.enconman.2019.01.103
6. H2TOOLS (2020) Hydrogen production. U.S. Department of Energy
7. Coskun Avci A, Toklu E (2021) A new analysis of two phase flow on hydrogen production from water electrolysis. Int J Hydrog Energy. https://doi.org/10.1016/j.ijhydene.2021.03.180.
8. Wu M, Fu Q, Huang J, Xu Q, Wang D, Liu X et al (2021) Effect of sodium dodecylbenzene sulfonate on hydrogen production from dark fermentation of waste activated sludge. Sci Total Environ 799:149383. https://doi.org/10.1016/j.scitotenv.2021.149383
9. Shi C, Elgarni M, Mahinpey N (2021) Process design and simulation study: CO_2 utilization through mixed reforming of methane for methanol synthesis. Chem Eng Sci 233:116364. https://doi.org/10.1016/j.ces.2020.116364
10. Bundhoo ZMA (2019) Potential of bio-hydrogen production from dark fermentation of crop residues: a review. Int J Hydro Energy 44:17346–62. https://doi.org/10.1016/j.ijhydene.2018.11.098
11. de Souza TAZ, Coronado CJR, Silveira JL, Pinto GM (2021) Economic assessment of hydrogen and electricity cogeneration through steam reforming-SOFC system in the Brazilian biodiesel industry. J Clean Prod 279:123814. https://doi.org/10.1016/j.jclepro.2020.123814

12. Mehmeti A, Angelis-Dimakis A, Arampatzis G, McPhail S, Ulgiati S (2018) Life cycle assessment and water footprint of hydrogen production methods: from conventional to emerging technologies. Environments 5:24. https://doi.org/10.3390/environments5020024

13. Wang C, Li L, Chen Y, Ge Z, Jin H (2021) Supercritical water gasification of wheat straw: Composition of reaction products and kinetic study. Energy 227:120449. https://doi.org/10.1016/j.energy.2021.120449

14. Siddiqui O, Dincer I (2021) Design and assessment of a new solar-based biomass gasification system for hydrogen, cooling, power and fresh water production utilizing rice husk biomass. Energy Convers Manag 236:114001. https://doi.org/10.1016/j.enconman.2021.114001

15. IEA (2019) The future of hydrogen. Futur Hydrog. https://doi.org/10.1787/1e0514c4-en

16. Boemke M (2021) The German hydrogen strategy. Watson Farley & Williams

17. Insights GM (2020) Hydrogen generation market to hit $160 billion by 2026. Says Global Market Insights, Inc., Selbyville, Dellaware

18. Al-Qahtani A, Parkinson B, Hellgardt K, Shah N, Guillen-Gosalbez G (2021) Uncovering the true cost of hydrogen production routes using life cycle monetisation. Appl Energy 281. https://doi.org/10.1016/J.APENERGY.2020.115958

19. Ritchie H, Roser M (2020) CO$_2$ and greenhouse gas emissions. OurWorldInDataOrg. https://ourworldindata.org/co2-and-other-greenhouse-gas-emissions

20. Espegren K, Damman S, Pisciella P, Graabak I, Tomasgard A (2021) The role of hydrogen in the transition from a petroleum economy to a low-carbon society. Int J Hydrogen Energy 46:23125–23138. https://doi.org/10.1016/J.IJHYDENE.2021.04.143

21. Navas-Anguita Z, García-Gusano D, Dufour J, Iribarren D (2021) Revisiting the role of steam methane reforming with CO$_2$ capture and storage for long-term hydrogen production. Sci Total Environ 771. https://doi.org/10.1016/J.SCITOTENV.2021.145432

22. Council H (2021) Hydrogen decarbonization pathways: a life-cycle assessment

23. Ali Khan MH, Daiyan R, Neal P, Haque N, MacGill I, Amal R (2021) A framework for assessing economics of blue hydrogen production from steam methane reforming using carbon capture storage & utilisation. Int J Hydrog Energy 46:22685–22706. https://doi.org/10.1016/J.IJHYDENE.2021.04.104

24. Moreira dos Santos R, Szklo A, Lucena AFP, de Miranda PE V (2021) Blue sky mining: strategy for a feasible transition in emerging countries from natural gas to hydrogen. Int J Hydrog Energy 46:25843–25859. https://doi.org/10.1016/J.IJHYDENE.2021.05.112

25. Boyano A, Blanco-Marigorta AM, Morosuk T, Tsatsaronis G (2011) Exergoenvironmental analysis of a steam methane reforming process for hydrogen production. Energy 36:2202–2214. https://doi.org/10.1016/j.energy.2010.05.020

26. Authayanun S, Arpornwichanop A, Patcharavorachot Y, Wiyaratn W, Assabumrungrat S (2011) Hydrogen production from glycerol steam reforming for low- and high-temperature PEMFCs. Int J Hydrog Energy 36:267–275. https://doi.org/10.1016/j.ijhydene.2010.10.061

27. Pashchenko D (2019) Thermochemical recuperation by ethanol steam reforming: thermodynamic analysis and heat balance. Int J Hydrog Energy 44:30865–30875. https://doi.org/10.1016/j.ijhydene.2019.10.009

28. Khila Z, Hajjaji N, Pons MN, Renaudin V, Houas A (2013) A comparative study on energetic and exergetic assessment of hydrogen production from bioethanol via steam reforming, partial oxidation and auto-thermal reforming processes. Fuel Process Technol 112:19–27. https://doi.org/10.1016/j.fuproc.2013.02.013

29. da SQ, Menezes JP, dos S, Dias AP, da Silva MAP, Souza MMVM (2020) Stability of Ni catalysts promoted with niobia for butanol steam reforming. Biomass Bioenergy 143:105882. https://doi.org/10.1016/j.biombioe.2020.105882

30. Pinto GM, de Souza TAZ, Coronado CJR, Flôres LFV, Chumpitaz GRA, Silva MH (2019) Experimental investigation of the performance and emissions of a diesel engine fuelled by blends containing diesel S10, pyrolysis oil from used tires and biodiesel from waste cooking oil 1–8. https://doi.org/10.1002/ep.13199

31. Mehrpooya M, Ghorbani B, Abedi H (2020) Biodiesel production integrated with glycerol steam reforming process, solid oxide fuel cell (SOFC) power plant. Energy Convers Manag 206. https://doi.org/10.1016/j.enconman.2020.112467

32. Yun J, Trinh N Van, Yu S (2021) Performance improvement of methanol steam reforming system with auxiliary heat recovery units. Int J Hydrog Energy 46:25284–25293. https://doi. org/10.1016/j.ijhydene.2021.05.032
33. Varkolu M, Kunamalla A, Jinnala SAK, Kumar P, Maity SK, Shee D (2021) Role of CeO2/ZrO2 mole ratio and nickel loading for steam reforming of n-butanol using Ni–CeO$_2$–ZrO$_2$–SiO$_2$ composite catalysts: a reaction mechanism. Int J Hydrog Energy 46:7320–7335. https://doi.org/10.1016/j.ijhydene.2020.11.240
34. Gao N, Salisu J, Quan C, Williams P (2021) Modified nickel-based catalysts for improved steam reforming of biomass tar: a critical review. Renew Sustain Energy Rev 145:111023. https://doi.org/10.1016/j.rser.2021.111023
35. Arregi A, Amutio M, Lopez G, Bilbao J, Olazar M (2018) Evaluation of thermochemical routes for hydrogen production from biomass: a review. Energy Convers Manag 165:696–719. https://doi.org/10.1016/j.enconman.2018.03.089
36. Jung S, Lee J, Moon DH, Kim K-H, Kwon EE (2021) Upgrading biogas into syngas through dry reforming. Renew Sustain Energy Rev 143:110949. https://doi.org/10.1016/j.rser.2021. 110949
37. Freitas ACD, Guirardello R (2014) Comparison of several glycerol reforming methods for hydrogen and syngas production using Gibbs energy minimization. Int J Hydrog Energy 39:17969–17984. https://doi.org/10.1016/j.ijhydene.2014.03.130
38. Aziz MAA, Setiabudi HD, Teh LP, Annuar NHR, Jalil AA (2019) A review of heterogeneous catalysts for syngas production via dry reforming. J Taiwan Inst Chem Eng 101:139–158. https://doi.org/10.1016/j.jtice.2019.04.047
39. Baruah R, Dixit M, Basarkar P, Parikh D, Bhargav A (2015) Advances in ethanol autothermal reforming. Renew Sustain Energy Rev 51:1345–1353. https://doi.org/10.1016/j.rser.2015. 07.060
40. Ahmed S, Krumpelt M (2001) Hydrogen from hydrocarbon fuels for fuel cells. Int J Hydrog Energy 26:291–301. https://doi.org/10.1016/S0360-3199(00)00097-5
41. Okolie JA, Nanda S, Dalai AK, Kozinski JA (2021) Techno-economic evaluation and sensitivity analysis of a conceptual design for supercritical water gasification of soybean straw to produce hydrogen. Bioresour Technol 331:125005. https://doi.org/10.1016/j.biortech.2021. 125005
42. Kumar S, Muthu Dineshkumar R, Ramanathan A (2021) Numerical prediction of gas composition from waste Ashoka and Neem leaves using downdraft gasifier. Mater Today Proc. https://doi.org/10.1016/j.matpr.2021.06.140
43. Bandara JC, Jaiswal R, Nielsen HK, Moldestad BME, Eikeland MS (2021) Air gasification of wood chips, wood pellets and grass pellets in a bubbling fluidized bed reactor. Energy 233:121149. https://doi.org/10.1016/j.energy.2021.121149
44. Bogdan VI, Koklin AE, Bogdan T V, Mishanin II, Kalenchuk AN, Laptinskaya TV et al (2020) Hydrogen generation by gasification of phenol and alcohols in supercritical water. Int J Hydrog Energy 45:30178–30187. https://doi.org/10.1016/j.ijhydene.2020.08.086
45. Leal AL, Soria MA, Madeira LM (2016) Autothermal reforming of impure glycerol for H2 production: thermodynamic study including in situ CO2 and/or H2 separation. Int J Hydrog Energy 41:2607–2620. https://doi.org/10.1016/j.ijhydene.2015.11.132
46. Thakkar M, Makwana JP, Mohanty P, Shah M, Singh V (2016) In bed catalytic tar reduction in the autothermal fluidized bed gasification of rice husk: Extraction of silica, energy and cost analysis. Ind Crops Prod 87:324–332. https://doi.org/10.1016/j.indcrop.2016.04.031
47. Demirbas A, Arin G (2002) An overview of biomass pyrolysis. Energy Sour 24:471–482. https://doi.org/10.1080/00908310252889979
48. Valliyappan T, Bakhshi NN, Dalai AK (2008) Pyrolysis of glycerol for the production of hydrogen or syn gas. Bioresour Technol 99:4476–4483. https://doi.org/10.1016/j.biortech. 2007.08.069
49. Gama Vieira GE, Pereira Nunes A, Fagundes Teixeira L, Nogueira Colen AG (2018) Biomassa: uma visão dos processos de pirólise. Rev Lib 15:167–178. https://doi.org/10.31514/ rliberato.2014v15n24.p167

50. Chumpitaz GRA, Coronado CJR, Carvalho JA, Andrade JC, Mendiburu AZ, Pinto GM et al (2019) Design and study of a pure tire pyrolysis oil (TPO) and blended with Brazilian diesel using Y-Jet atomizer. J Brazilian Soc Mech Sci Eng 41:139. https://doi.org/10.1007/s40430-019-1632-z

51. Di Stasi C, Cortese M, Greco G, Renda S, González B, Palma V et al (2021) Optimization of the operating conditions for steam reforming of slow pyrolysis oil over an activated biochar-supported Ni–Co catalyst. Int J Hydrog Energy 46:26915–26929. https://doi.org/10.1016/j.ijhydene.2021.05.193

52. Xu B, Argyle MD, Shi X, Goroncy AK, Rony AH, Tan G et al (2020) Effects of mixture of CO_2/CH_4 as pyrolysis atmosphere on pine wood pyrolysis products. Renew Energy 162:1243–1254. https://doi.org/10.1016/j.renene.2020.08.069

53. Ahmad MS, Klemeš JJ, Alhumade H, Elkamel A, Mahmood A, Shen B et al (2021) Thermo-kinetic study to elucidate the bioenergy potential of Maple Leaf Waste (MLW) by pyrolysis, TGA and kinetic modelling. Fuel 293:120349. https://doi.org/10.1016/j.fuel.2021.120349

54. Kumar RS, Sivakumar S, Joshuva A, Deenadayalan G, Vishnuvardhan R (2021) Bio-fuel production from Martynia annua L. seeds using slow pyrolysis reactor and its effects on diesel engine performance, combustion and emission characteristics. Energy 217:119327. https://doi.org/10.1016/j.energy.2020.119327

55. Reginatto V, Antônio RV (2015) Fermentative hydrogen production from agroindustrial ligno-cellulosic substrates. Braz J Microbiol 46:323–335. https://doi.org/10.1590/S1517-838246220140111

56. Akhlaghi N, Najafpour-darzi G (2020) A comprehensive review on biological hydrogen production. Int J Hydrog Energy 45:22492–22512. https://doi.org/10.1016/j.ijhydene.2020.06.182

57. Pan S, Alex M, Huang I, Liu I, Chang E, Chiang P (2015) Strategies on implementation of waste-to-energy (WTE) supply chain for circular economy system: a review. J Clean Prod 108:409–421. https://doi.org/10.1016/j.jclepro.2015.06.124

58. Das D, Veziroglu TN (2008) Advances in biological hydrogen production processes. Int J Hydrog Energy 33:6046–6057. https://doi.org/10.1016/j.ijhydene.2008.07.098

59. Li C, Fang HHP (2007) fermentative hydrogen production from wastewater and solid wastes by mixed cultures. Crit Rev Environ Sci Technol 37:1–39. https://doi.org/10.1080/10643380600729071

60. Manish S, Banerjee R (2008) Comparison of biohydrogen production processes. Int J Hydrog Energy 33:279–286. https://doi.org/10.1016/j.ijhydene.2007.07.026

61. Mahata C, Ray S, Das D (2020) Optimization of dark fermentative hydrogen production from organic wastes using acidogenic mixed consortia. Energy Convers Manag 219. https://doi.org/10.1016/j.enconman.2020.113047

62. Ntaikou I, Antonopoulou G, Lyberatos G (2010) Biohydrogen production from biomass and wastes via dark fermentation: a review. Waste Biomass Valorization 21–39. https://doi.org/10.1007/s12649-009-9001-2

63. Sikora A, Blaszczyk M, Jurkowski M, Zielenkiewicz U (2013) Lactic acid bacteria in hydrogen-producing consortia: on purpose or by coincidence? Lact Acid Bact RD Food, Heal Livest Purp InTech https://doi.org/10.5772/50364

64. Anukam A, Mohammadi A, Naqvi M, Granström K (2019) A review of the chemistry of anaerobic digestion: methods of accelerating and optimizing process efficiency. Processes 7:1–19

65. Lukajtis R, Holowacz I, Kucharska K, Glinka M, Rybarczyk P, Przyjazny A et al (2018) Hydrogen production from biomass using dark fermentation. Renew Sustain Energy Rev 91:665–694. https://doi.org/10.1016/j.rser.2018.04.043

66. De Gioannis G, Muntoni A, Polettini A, Pomi R (2013) A review of dark fermenta-tive hydrogen production from biodegradable municipal waste fractions. Waste Manag 33:1345–1361. https://doi.org/10.1016/j.wasman.2013.02.019

67. Sikora A, Detman A, Chojnacka A, Blaszczyk MK (2017) Anaerobic digestion: I. A common process ensuring energy flow and the circulation of matter in ecosystems. II. A tool for the

production of gaseous biofuels world's largest science. Technology & Medicine Open Access book publisher. Ferment. Process. 1ᵃ, pp 271–301. https://doi.org/10.5772/64645

68. Thauer RK, Jungermann K, Decker K (1977) Energy conservation in chemotrophic anaerobic bacteria. Bacteriol Rev 41:100–180

69. Hallenbeck PC (2005) Fundamentals of the fermentative production of hydrogen. Water Sci Technol 52:21–29

70. Guo XM, Latrille E, Carrère H, Steyer J-P, Trably E (2010) Hydrogen production from agricultural waste by dark fermentation: a review. Int J Hydrog Energy 35:10660–10673. https://doi.org/10.1016/j.ijhydene.2010.03.008

71. Kamaraj M, Ramachandran KK, Aravind J (2020) Biohydrogen production from waste materials: benefits and challenges. Int J Environ Sci Technol 17:559–576. https://doi.org/10.1007/s13762-019-02577-z

72. Sawatdeenarunat C, Surendra KC, Takara D, Oechsner H, Kumar S (2015) Anaerobic digestion of lignocellulosic biomass: challenges and opportunities. Bioresour Technol 178:178–186. https://doi.org/10.1016/j.biortech.2014.09.103

73. Freitas LC, Barbosa JR, da Costa ALC, Bezerra FWF, Pinto RHH, de Carvalho Junior RN (2021) From waste to sustainable industry: how can agro-industrial wastes help in the development of new products? Resour Conserv Recycl 169. https://doi.org/10.1016/j.resconrec.2021.105466

74. Shrestha S, Fonoll X, Kumar S, Raskin L (2017) Biological strategies for enhanced hydrolysis of lignocellulosic biomass during anaerobic digestion: Current status and future perspectives. Bioresour Technol 245:1245–1257. https://doi.org/10.1016/j.biortech.2017.08.089

75. Wang S, Dai G, Yang H, Luo Z (2017) Lignocellulosic biomass pyrolysis mechanism: a state-of-the-art review. Prog Energy Combust Sci 62:33–86. https://doi.org/10.1016/j.pecs.2017.05.004

76. Bhatia SKS, Jagtap S, Bedekar AA, Bhatia RK, Rajendran K, Pugazhendhi A et al (2021) Renewable biohydrogen production from lignocellulosic biomass using fermentation and integration of systems with other energy generation technologies. Sci Total Environ 765. https://doi.org/10.1016/j.scitotenv.2020.144429

77. Bundhoo MAZ, Mohee R, Hassan MA (2015) Effects of pre-treatment technologies on dark fermentative biohydrogen production: A review. J Environ Manage 157:20–48. https://doi.org/10.1016/j.jenvman.2015.04.006

78. dos Santos AC, Ximenes E, Kim Y, Ladisch MR (2019) Lignin-Enzyme Interactions in the Hydrolysis of Lignocellulosic Biomass. Trends Biotechnol 37:518–531. https://doi.org/10.1016/j.tibtech.2018.10.010

79. Baruah J, Nath BK, Sharma R, Kumar S, Deka RC, Baruah DC et al (2018) Recent trends in the pretreatment of lignocellulosic biomass for value-added products. Front Energy Res 6:1–19. https://doi.org/10.3389/fenrg.2018.00141

80. El-bery H, Tawfik A, Kumari S, Bux F (2013) Effect of thermal pre-treatment on inoculum sludge to enhance bio-hydrogen production from alkali hydrolysed rice straw in a mesophilic anaerobic baffled reactor. Environ Technol 37–41. https://doi.org/10.1080/09593330.2013.824013

81. Mirza SS, Qazi JI, Chen YL (2019) Growth characteristics and photofermentative biohydrogen production potential of purple non sulfur bacteria from sugar cane bagasse. Fuel 255. https://doi.org/10.1016/j.fuel.2019.115805

82. Jiang D, Ge X, Zhang T, Liu H (2016) Photo-fermentative hydrogen production from enzymatic hydrolysate of corn stalk pith with a photosynthetic consortium. Int J Hydrog Energy 41:16778–16785. https://doi.org/10.1016/j.ijhydene.2016.07.129

83. Datar R, Huang J, Maness P, Mohagheghi A, Czernik S, Chornet E (2007) Hydrogen production from the fermentation of corn stover biomass pretreated with a steam-explosion process. Int J Hydrog Energy 32:932–939. https://doi.org/10.1016/j.ijhydene.2006.09.027

84. Cao G, Ren N, Wang A, Guo W, Xu J, Liu B (2010) Effect of lignocellulose-derived inhibitors on growth and hydrogen production by thermoanaerobacterium thermosaccharolyticum W16. Int J Hydrog Energy 35:13475–13480. https://doi.org/10.1016/j.ijhydene.2009.11.127

85. Zagrodnik R, Duber A, Seifert K (2021) Hydrogen production during direct cellulose fermentation by mixed bacterial culture: the relationship between the key process parameters using response surface methodology. J Clean Prod 314. https://doi.org/10.1016/j.jclepro.2021.127971

86. Bahry H, Abdallah R, Chezeau B, Pons A, Taha S (2019) Biohydrogen production from carob waste of the Lebanese industry by dark fermentation fermentation. Biofuels 1–11. https://doi.org/10.1080/17597269.2019.1669862

87. Rödl A, Wulf C, Kaltschmitt M (2018) Assessment of selected hydrogen supply chains—factors determining the overall GHG emissions. Hydrog Supply Chain Des Deploy Oper 81–109. https://doi.org/10.1016/B978-0-12-811197-0.00003-8

88. Oliveira AM, Beswick RR, Yan Y (2021) A green hydrogen economy for a renewable energy society. Curr Opin Chem Eng 33. https://doi.org/10.1016/J.COCHE.2021.100701

89. IRENA (2018) Hydrogen from renewable power: technology outlook for the energy transition

90. Nechache A, Hody S (2021) Alternative and innovative solid oxide electrolysis cell materials: a short review. Renew Sustain Energy Rev 149. https://doi.org/10.1016/J.RSER.2021.111322

91. Mohammadi A, Mehrpooya M (2018) Techno-economic analysis of hydrogen production by solid oxide electrolyzer coupled with dish collector. Energy Convers Manag 173:167–178. https://doi.org/10.1016/J.ENCONMAN.2018.07.073

92. Wulf C, Linssen J, Zapp P (2018) Power-to-gas—concepts, demonstration, and prospects. Hydrog Supply Chain. Academic Press, pp 309–345. https://doi.org/10.1016/B978-0-12-811197-0.00009-9

93. Ozturk M, Dincer I (2021) A comprehensive review on power-to-gas with hydrogen options for cleaner applications. Int J Hydrog Energy 46:31511–31522. https://doi.org/10.1016/J.IJHYDENE.2021.07.066

94. Yu M, Wang K, Vredenburg H (2021) Insights into low-carbon hydrogen production methods: green, blue and aqua hydrogen. Int J Hydrog Energy 46:21261–21273. https://doi.org/10.1016/J.IJHYDENE.2021.04.016

95. Bhandari R, Trudewind CA, Zapp P (2014) Life cycle assessment of hydrogen production via electrolysis—a review. J Clean Prod 85:151–163. https://doi.org/10.1016/J.JCLEPRO.2013.07.048

96. Singh V, Dincer I, Rosen MA (2018) Life cycle assessment of ammonia production methods. Exergetic, Energ Environ Dimens 935–959. https://doi.org/10.1016/B978-0-12-813734-5.00053-6

97. Julio AAV, Ocampo Batlle EA, Trindade AB, Nebra SA, Reyes AMM, Escobar Palacio JC (2021) Energy, exergy, exergoeconomic, and environmental assessment of different technologies in the production of bio-jet fuel by palm oil biorefineries. Energy Convers Manag 243. https://doi.org/10.1016/j.enconman.2021.114393

98. Liu H, Liu S (2021) Life cycle energy consumption and GHG emissions of hydrogen production from underground coal gasification in comparison with surface coal gasification. Int J Hydrog Energy 46:9630–9643. https://doi.org/10.1016/J.IJHYDENE.2020.12.096

99. Candelaresi D, Valente A, Iribarren D, Dufour J, Spazzafumo G (2021) Comparative life cycle assessment of hydrogen-fuelled passenger cars. Int J Hydrog Energy. https://doi.org/10.1016/J.IJHYDENE.2021.01.034

100. Lotric A, Sekavcnik M, Kuštrin I, Mori M (2021) Life-cycle assessment of hydrogen technologies with the focus on EU critical raw materials and end-of-life strategies. Int J Hydrog Energy 46:10143–10160. https://doi.org/10.1016/J.IJHYDENE.2020.06.190

101. Howarth RW, Jacobson MZ (n/a) How green is blue hydrogen? Energy Sci Eng. https://doi.org/10.1002/ese3.956

102. Cetinkaya E, Dincer I, Naterer GF (2012) Life cycle assessment of various hydrogen production methods. Int J Hydrog Energy 37:2071–2080. https://doi.org/10.1016/j.ijhydene.2011.10.064

103. Rau GH, Willauer HD, Ren ZJ (2018) The global potential for converting renewable electricity to negative-CO_2-emissions hydrogen. Nat Clim Chang 8:621–625. https://doi.org/10.1038/s41558-018-0203-0

104. Valente A, Iribarren D, Dufour J (2021) Comparative life cycle sustainability assessment of renewable and conventional hydrogen. Sci Total Environ 756. https://doi.org/10.1016/J.SCI TOTENV.2020.144132

105. Battista F, Montenegro Camacho YS, Hernández S, Bensaid S, Herrmann A, Krause H et al (2017) LCA evaluation for the hydrogen production from biogas through the innovative BioRobur project concept. Int J Hydrog Energy 42:14030–14043. https://doi.org/10.1016/j.ijhydene.2016.12.065

106. Institute GC (2020) Global status of CCS 2020

107. IEA (2020) Hydrogen. Paris

108. McKenna RC, Bchini Q, Weinand JM, Michaelis J, König S, Köppel W et al (2018) The future role of power-to-gas in the energy transition: regional and local techno-economic analyses in Baden-Württemberg. Appl Energy 212:386–400. https://doi.org/10.1016/J.APENERGY.2017.12.017

109. Jaworski J, Kulaga P, Blacharski T (2020) Study of the effect of addition of hydrogen to natural gas on diaphragm gas meters. Energies 13. https://doi.org/10.3390/en13113006

110. Ince AC, Colpan CO, Hagen A, Serincan MF (2021) Modeling and simulation of power-to-x systems: a review. Fuel 304. https://doi.org/10.1016/J.FUEL.2021.121354

111. Australia G of S (2020) Hydrogen in the gas distribution networks. Clean Energy Transit Dep Energy Min 149

112. Ferreira TVB, Machado GV, Soares JB, Achão C, Almeida EM de, Botelho GML et al (2021) Bases para a Consolidação da Estratégia Brasileira do Hidrogênio. Nota Técnica, Ministério Minas e Energia Bras 34

113. Silveira JL (2017) Sustainable hydrogen production processes. Springer International Publishing. https://doi.org/10.1007/978-3-319-41616-8

114. Paulino RFS, Silveira JL (2019) Chapter 4—Biohydrogen production and bagasse gasification process in the sugarcane industry. In: Rai M, Ingle AP (eds) Sustain. Bioenergy, Elsevier, pp 89–126. https://doi.org/10.1016/B978-0-12-817654-2.00004-6

115. Silveira JL, Tuna CE, Pedroso DT, da Silva ME, Machin EB, Braga LB et al (2014) Technological advancements in biohydrogen production and bagasse gasification process in the sugarcane industry with regard to Brazilian conditions. In: da Silva SS, Chandel AK (eds) Biofuels Brazil fundam. Asp Recent Dev Futur Perspect. Springer International Publishing, Cham, pp 393–411. https://doi.org/10.1007/978-3-319-05020-1_18

Designing an Energy Use Analysis and Life Cycle Assessment of the Environmental Sustainability of Conservation Agriculture Wheat Farming in Bangladesh

Md Mashiur Rahman⑩, Md Sumon Miah, Md Aminur Rahman, Mukaddasul Islam Riad, Naznin Sultana, Monira Yasmin, Fouzia Sultana Shikha, and Md Manjurul Kadir

Abstract The agricultural sector in Bangladesh is an ongoing societal expectation of reducing environmental impacts and increasing crop productivity to provide food security for its growing population. Introducing life cycle assessment is a systematic approach for establishing how sustainable a crop may become and the potential impacts of complete life cycle wheat farming on the environment and input resource conservation. This innovative field study focuses on conservation agriculture wheat farming to increase energy use efficiency (EUE) and environmental sustainability by decreasing greenhouse gas (GHG) emissions through comparing different conservation tillage practices to conventional tillage. Furthermore, the study estimated the net

M. M. Rahman (✉)
Agricultural Engineering Division, Regional Agricultural Research Station, Bangladesh Agricultural Research Institute, Jamalpur 2000, Bangladesh

M. S. Miah
Scientific Officer, Farm Machinery and Postharvest Process Engineering Division, Bangladesh Agricultural Research Institute, Gazipur 1701, Bangladesh

M. A. Rahman
Agricultural Engineering Division, Regional Agricultural Research Station, Bangladesh Agricultural Research Institute, Rahmatpur, Barisal, Bangladesh

M. I. Riad
Farm Division, Regional Agricultural Research Station, Bangladesh Agricultural Research Institute, Jamalpur 2000, Bangladesh

N. Sultana
Entomology Division, Bangladesh Sugarcrop Research Institute, Ishwardi, Pabna 6620, Bangladesh

M. Yasmin · F. S. Shikha
Soil Science Division, Regional Agricultural Research Station, Bangladesh Agricultural Research Institute, Jamalpur 2000, Bangladesh

M. M. Kadir
Chief Scientific Officer, Regional Agricultural Research Station, Bangladesh Agricultural Research Institute, Jamalpur 2000, Bangladesh

© The Author(s), under exclusive license to Springer Nature Singapore Pte Ltd. 2022
S. S. Muthu (ed.), *Environmental Footprints of Crops*,
Environmental Footprints and Eco-design of Products and Processes,
https://doi.org/10.1007/978-981-19-0534-6_5

carbon footprint (CF) of wheat farming, taking into account the additional contribution of soil carbon sequestration and offered a model of the environmental sustainability for wheat farming. The FEAT tool was used to assess energy use analysis, life cycle GHG emissions, and CF during the life cycle (CF) wheat farming. The introduced strip tillage (ST), minimum tillage (MT), and conventional tillage (CT) were predicted by utilizing input enegy of 18,764.29, 18,728.78, and 20,564.32 MJ ha^{-1} in wheat farming, respectively, with the EUE of 8.46, 8.65, and 6.25%. Among the tillage practices, MT is the most effective practice option in the wheat farming production process. The net life cycle GHG emissions were observed to be 1.968, 1.977, and 2.023 kgCO$_2$eq ha^{-1} for ST, MT and CT, respectively, where the CF was estimated to be 0.013, 0.012, and 0.014 kgCO$_2$ MJ^{-1}. As a result, CA-based ST and MT practices to be the most effective life cycle GHG mitigation options for wheat farming in Bangladesh.

Keywords Conservation Agriculture · Conservation tillage · Life cycle assessment · Energy footprint · Greenhouse gas emission · Carbon footprint

Highlights

1. Strip tillage (ST), Minimum tillage (MT), and conventional tillage (CT) were introduced.
2. Conservation tillage affected carbon footprint through soil organic carbon accumulation and yield.
3. MT practice had the lowest energy input calculated in 18,728.78 MJha^{-1}.
4. Minimum net life cycle GHG emissions was 1.968 kgCO$_2$eqha^{-1} for ST practice.
5. The carbon footprint was estimated as 0.012 kgCO$_2$MJ^{-1} of energy output for MT practice.

1 Introduction

The agriculture sector in Bangladesh is expecting to increase the crop production through a sustainable management practices for overcoming ongoing increasing food security due to growing its population of 164.7 million currently [72] and awaited to reach 205 million by 2050 [10], while need to reduce energy inputs to secure agriculture profitability and also ensure environmental sustainability in term of greenhouse gas (GHG) emissions [11, 58, 60]. Emerging conservation agriculture (CA) can be suggested as one of the solutions in sustainable crop management practices which consisting of minimal soil disturbance, crop residue retention, and diverse crop rotations for the possibility to obtain higher crop productivity, increasing energy use efficiency (EUE), improving soil quality and also reducing GHG emissions [17, 18, 39, 79], which is practiced over 3.9 Mha area of South Asia [38]. Being implemented CA practice through management of minimal soil disturbance using conservation tillage practices (CTP) and covering stable crop residue has grown in importance seeking to improve EUE and reduce life cycle GHG emissions of agricultural farming, as well as increasing crop productivity [30, 69, 70]. These benefits have identical advantages to increasing adoption of CA practices worldwide [73, 77].

CA-based CTP is a tillage system of minimizing soil disturbance which creates a suitable soil environment for growing crops by way of conserving soil, water, and energy resources, mainly through reduction of the intensity of tillage passes and a minimum of 30% of the topsoil covered with the stable crop residues [61]. With this, CTP has a significant influence on soil physical and chemical properties for maintaining soil health due to tilling soil with a minimal; other main benefits of CTP are including—reduce soil erosion, decreased labor and energy inputs, increased availability of water for crop production [20]. Moreover, covering stable crop residue also benefits a sustainable crop production system that increases soil organic carbon (SOC), controls weeds, and improves soil structure that helps to reduce energy inputs, anchoring soil and adding carbon deep in the soil profile via roots [20].

Different tillage practices, including CTP such as minimum tillage (MT), strip tillage (ST), and conventional tillage (CT), are such crop management options based on the principles of CA [53], and this has been identified for mitigating GHG emissions in wheat farming [1, 12, 26, 37, 44, 52, 53]. Among these, ST and MT might be a good options for reducing GHG emissions and enhancing the SOC in the top layer of soil [21]. By adopting CTPs, GHG emissions can be reduced significantly in wheat farming by increasing SOC accumulation in soil [11]. On the contrary, soil tillage practices are an essential part of the agriculture production process, which has adverse effects on physicochemical properties of soil and environmental impacts [4]. Nonetheless, immense tillage practice has led to higher aeration that causes SOC breakdown and also increases GHG emissions [5]. The researcher reported that the CT and reduced tillage affected soil losses of about 207.7 and 111.5 Mg ha^{-1} [6].

Energy input is a crucial parameter used in the agricultural sector for producing food grains, although it is one of the main contributors to adverse environmental impacts for emitting GHG emissions [34, 76]. Agriculture has an energy use of 5% share globally [9]. Among these, the most energy sources come from non-renewable sources, leading to enormous GHG emissions. Various energy sources are utilized in the agriculture sector like human, renewable, coal, fossil fuel, solar, wind, and hydro, etc. Among these, fossil (diesel) fuel is used in higher amounts in developing countries for different machinery uses in agricultural farming operations [56]. Higher energy inputs, especially non-renewable energy inputs, reduce EUE in agriculture farming. In addition, many indirect energy inputs are used in agriculture farming, such as mineral or synthesis fertilizers, chemical pesticides, insecticides, and herbicides. For this, current energy consumptions in the agricultural production process need to be minimized for undertaking use of optimum level, which is the first assumption seeking to optimize any farming activities in terms of production cost to gain farm profitability and achieve agricultural sustainability [45, 62]. Therefore, analysis of energy use needs to be assessed to determine the EUE and environmental impacts for achieving the energy footprint of wheat farming. According to the Paris agreement on the climate change management strategy, the primary emphasis has been given on reducing GHG emissions. Therefore, the ongoing agriculture farming requires to turn into the practice by adopting climate-smart agriculture. Hence, the different environmental footprints such as carbon and energy footprint are needed to assess environmental sustainability of crops.

However, recent agriculture and agricultural practices in Bangladesh have emitted GHG emissions of 37% among all sectors, where the combined share of agriculture and agricultural land use emitted 47.2%. In contrast, world agriculture and related land use contribute 19.79% of GHG emissions [24]. United Nations Environment Programme (UNEP) emissions gap report revealed that about one-third of the GHG emissions attributed is from agriculture and related land use [67]. It also reported that agriculture is one of the main four sectors contributing to GHG emissions and has proven to reduce GHG emissions. The report underlined that promoting CA practices in agriculture should play the proper role in reducing GHG emissions. That is why there is still hope for minimizing GHG emissions from the agriculture sector by applying conservation tillage practices [24, 58]. The agricultural sector has contributed to emitting more GHG concentrations into the atmosphere in recent decades. Improper soil management and the use of high amounts of energy (direct and indirect) inputs contribute significantly to these emissions [21].

All about it, life cycle GHG emissions is one of the global-scale climate change indicators to determine the environmental sustainability of any crop farming [29]. The amount of life cycle GHG emissions of crop or food production or services product is commonly expressed as carbon footprint (CF) [31, 54], where CF is a sustainability indicator to assess global warming potential (GWP) with a solid scientific basis, and it is measured as of carbon dioxide equivalents (CO_2eq) [71]. Recently, life cycle assessment (LCA) is the most employed approach to quantify different environmental footprints on life cycle farming activities. To assess the environmental impacts, use the term CF using the amount of GHG emissions per kilogram/ton of food or grains produced throughout the entire production process in all stages of a crop like tillage operation, application of fertilizers and pesticides, harvesting, processing, transport, storage, and consumption in the end during its life cycle [14]. Data of GHG emissions can be measured with the field measurement instruments or calculations based on emission factors (EF) given by IPCC [46]. The application of LCA analysis in wheat farming is not a new technolgoy and several international studies have investigated its importance as a user-friendly tool to evaluate sustainability within agricultural farming. In this study, the LCA approach was employed to assess the potential environmental footprint of wheat farming associated with the combination of different tillage practices, including CT practices and synthetic fertilizer, which was not investigated in the past.

Regard as mentioned above, this field study has been undertaken which has a special significance. In most of the studies, a higher share of energy inputs comes from agricultural machinery and synthetic chemical fertilizers, and the inputs energy consumption is also differed by different tillage practices. CTP, including ST and MT, is related to the lower fuels and energy inputs. In contrast, a CT practice is coupled with higher fuels and energy inputs. For this, a fundamental problem is the impact assessment of different tillage practices on EUE and different environmental footprints in agroecosystems. Therefore, considering preliminary study in agroecosystem of conservation tillage wheat farming, this field experiment was conducted for the different tillage practices associated with the stable crop residue retention (20 cm) to evaluate the EUE, life cycle GHG emissions, and its impacts on the environmental footprint including energy and carbon footprint. In addition, this study investigated the performance comparison of energy use analysis and environmental footprint for the different tillage practices (ST, MT, and CT). Finally, a systematic LCA approach proposed a suitable model for sustainable wheat farming.

2 Materials and Methods

2.1 Study Site, Design, and Soil Sampling

The conducted field study was located in the research field at the Regional Agriculture Research Station (RARS), Jamalpur, Bangladesh, in the rabi season during 2019–21 and the site details are shown in Table 1. The experimental soil is in agro-ecological zone (AEZ) 8 and 9 of the young & old Brahmaputra and Jamuna floodplain of Bangladesh. The climatic condition was represented by semi-arid monsoon and subtropical within a variation of rainfall during the entire year. The crop was Wheat (*Triticum aestivum* L.)—BARI Gom 28 used in wheat farming. The unit area of experimental plots was 15 m × 13 m alongside a 2-m buffer distance among the experimental plots. The field experiment was performed in the same management practices followed by the same layout. The design of the experiment was adopted by a randomized complete block (RCB) design with four replications. The study treatments were based on different tillage practices and stable crop residue retention (20 cm), shown in Table 2. At the starting of the experiment, each plot was divided from the other according to the layout. The respective tillage practices were done by tillage machinery and stable crop residue retention was maintained with the previous crop of rice used in farmer's practice.

Table 1 Characteristics of the study field in wheat farming

Parameters	Details
Location	Regional Agricultural Research Station (RARS), Jamalpur, Bangladesh
Soil type	Silt clay loam
Location	24°56′32.3″ N latitude, 89°55′37.8″ E longitude, and altitude of 16.46 m
Rainfall	1549.45 mm, medium level Average rainfall (during November–March), 440 mm, concentrated in monsoon season (June–September)
Drainage	Moderate
Temperature	Maximum, 32 °C and minimum, 20 °C (avg.)

Table 2 Experimental treatment of wheat farming

Crop	Conservation tillage practices	Residue management
Wheat	T_1 = Strip tillage (ST), no disturbances of the soil	Stable crop residue management (20 cm)
	T_2 = Minimum Tillage (MT); single tillage	
	T_3 = Conventional Tillage (CT), farmer's practice	

At the pre-sowing and post-harvesting stages, soil samples from 0 to 20 cm deep were collected employing a soil auger. Nine soil samples from each treatment were collected, and the physicochemical properties of soils were determined the procedure delineated by the United States Department of Agriculture (USDA), National Soil Survey Center [14].

Table 3 shows the findings of the soil property study prior to land preparation. Soil carbon sequestration is a continuous process in soil activities to accumulate the soil organic carbon (SOC), which is applied to estimate carbon accounting, as the amount of SOC accumulation for CA formation during the initial soil condition and after crop harvesting differs. The SOC content in soil was determined using the wet oxidation technique. [36].

2.2 Soil Tillage and Agronomic Management Practices

Figure 1 shows the flowsheet of the wheat production process with the account to the LCA system boundary. Five days before planting, tillage practice was done in different treatments according to the design. The previous stable rice crop residue was left on the soil (20 cm) and pulverized during the tillage process. A power tiller-operated BARI inclined seeder was employed for land preparation and seed sowing in both strip tillage (ST) and minimum tillage (MT) practices [32]. For ST practice, the tine setting of the BARI inclined seeder was modified to 12 tines for tilling purposes at each journey and seed sowing was done in six rows with a 20 cm line to line spacing. Thus, the soil was ploughed to a depth of 6–7 cm, and the wheat seed was sown in lines at a specific distance using a seed metering device at a seed rate of 120 kg ha^{-1}. The seed rate was calibrated following the BARI's standard seed rate [19]. One machine operator and one labor were only needed to perform tilling and sowing operations. Soil tilling practice was maintained to a minimum level under the MT practice. The primary goal of MT pracitce is to pulverize the soil, along with the crop residue and weeds. This method was used to incorporate crop residue and weeds into the topsoil, increasing the SOC in the soil. A power tiller-operated BARI inclined seeder with 48 tines was used to till the soil surface, and the seed planting operation was carried out, using the same approach as ST practice. The only difference between these two tillage methods is that tillage practice; ST was followed for tilling in lines

Table 3 Initial soil analysis results in wheat farming

(a) Physical characteristics of soil

Soil depth (cm)	Particle density (g/cm³)	Bulk density (g/cm³)	Porosity (%)	Infiltration (mm/hr)	Field capacity (%)	Soil texture
0–20	2.62	1.40	46.81	8.50	29.41	Silt clay loam

(b) Chemical characteristics of soil

Sample	pH	SOM (%)	SOC (%)	Ca	Mg	K	Total N (%)	P	S	B	Cu	Fe	Mn	Zn
				m_{eq} 100 g^{-1}				μg g^{-1}						
Mean	6.5	2	1.16	6	1.9	0.12	0.1	4.81	8.76	0.32	2.4	24	4	1.22
Critical level	–	–	–	2	0.5	0.12	–	10	10	0.2	0.2	4	1	0.6

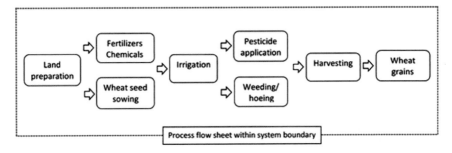

Fig. 1 Wheat production flow sheet in the LCA system boundary

where line outside soils were untilled, and MT was followed for tilling all over the soil up to 6–7 cm depth. One operator and one labor were needed to perform tilling and sowing operations in MT practice. Both ST and MT practices can save a significant amount of time and fuel compared to the conventional tillage (CT) practice [60]. CT practice, known as commonly farmer tillage practice, is used for intensive operations in Bangladesh. Tillage was performed with a power tiller up to 16 cm, pulverizing the soil 4–5 times, while removing exposed weeds from the topsoil. The seed sowing operation was performed manually by broadcasting. This operation took a long time and more fuel was consumed in the tilling operation, resulting in higher energy inputs. In these circumstances, an adverse soil environment was created when higher soil tillage was performed for land preparation, which was unfavorable for growing the beneficial soil microorganisms that are responsible for biomass decomposition and recirculation of biogenic elements that makes nutrients available to plants and growth of SOC concentration [27].

The application rate of chemical fertilizers for the first year was $N_{120}P_{108}K_{80}S_{85}Zn_{10}B_{5.2}$ based on the initial soil test in order to higher yield goal [7] and for the second year, the fertilizer application rate was $N_{60}P_{52}K_{40}S_{85}Zn_{10}B_{5.2}$ to minimize the environmental effects. At land preparation time, urea (one-third) was mixed with triple superphosphate (TSP), MoP, gypsum, and $ZnSO_4$ and then mixed fertilizers was applied to the soil. Half of the urea from the remaining quantity was applied at 25 days after sowing (DAS) and the remaining urea was applied at 45 DAS. Irrigation was used after each fertilizer application. When necessary, the weeder machinery performed intercultural operations. Weeds in ST plots were partially controlled by spraying a post-emergence selective herbicide, Affinity (Carfentrazone ethyl + Isoproturon) @ 2.5 g/L water at 25 DAS, with just one hand weeding at 28 DAS for complete weed removal. The gravimetric method was used to closely monitor the soil moisture[13]. Wheat seeds were sown in the study field on 15 November 2019. During the whole duration, irrigation activities were carried out three times in a volumetric and regulated manner by the measurement system. Irrigation water was applied based on growth stages and DAS; first irrigation was applied after sowing to supply available moisture in the soil, creating favorable conditions for seed germination; second irrigation at 25 DAS was applied when milk stages emerged; and final irrigation was applied at days of heading (51 DAS). An herbicide spray was applied at 25 DAS on the same day of irrigation applied. Harvesting operation of wheat farming was done by a combined harvester of 16 hp. Fuel measurement was done during farming activities while taking into account time losses for operator personnel [2]. The following equation (1) was used to compute the production energy for agricultural machinery and harvesters [16]:

$$M_e = GM_{pe}/TW \tag{1}$$

where M_e represents the energy of machine per unit area (MJ/ha^{-1}), G represents the machine mass (kg), M_{pe} represents the machine production energy (MJ kg^{-1}), T represents the economic life (h), and W represents the effective field capacity (hah^{-1}).

2.3 LCA Modeling

The LCA is a useful tool for evaluating the environmental sustainability of a process, product, or system over the course of its entire life cycle [68], which was applied in this study to determine the life cycle GHG emissions as a global warming potential (GWP). This study of wheat farming took into account from cradle to farmgate for considering an LCA analysis of energy and carbon footprint [35]. A farm energy analysis tool (FEAT) was employed as a whole-farm approach within the LCA system boundary to estimate energy and carbon footprint, where FEAT is a static, deterministic, and database tool. The newest version developed in 2018 was used in this analysis, together with the most suitable historical database for this study [15]. This database was selected as the reference library to complete the inventory analysis as it is mainly used to help the agricultural sector improve its practices. An advantage of using the FEAT tool over others is that it can be assimilated with it for its transparency and accuracy of any local data. This program is simple to use, allowing it to be used as a dynamic crop production tool to study life cycle energy usage and GHG emissions. If emission factors (EF) were not accessible for any inputs, they were created using a combination of generic and local data from Bangladesh.

2.3.1 Modeling of Life Cycle Energy Flow

Life cycle energy flow modeling is necessary to determine the optimum use of energy inputs for respective agricultural crop and their impact on improving the EUE from the farming inputs. Based on energy consumption, agriculture input energy is separated into direct and indirect energy [76]. Machinery used in agricultural operations, such as tillage for land preparation, seed sowing, fertilization, irrigation, harvesting, threshing, and transportation, were performed by different agricultural machinery, are known as direct input energy. Human power was adopted in every operation of wheat farming in Bangladesh. Besides these, other necessary inputs energy was also applied in wheat farming which is called indirect input energy, such as farm machinery, seed, pesticide, and chemical fertilizer. For assessing optimum energy usage in wheat farming, energy balance calculations employing direct (operational) and indirect energy were examined. Many other types of energy, such as renewable energy such as seed and human power, and non-renewable energy such as chemicals, fertilizer, herbicides, diesel fuel, water, and machinery, were also identified.

Life cycle energy input–output usage was determined using different energy inputs and outputs, wheat grain yield, and biomass (wheat residue) output (Table 4). Energy equivalents of different inputs and outputs parameters for conservation agriculture wheat farming were extracted from related studies (Table 5). After then, input and output energy of CA-based wheat farming for different tillage practices were calculated by their respective quantity multiplying with their energy equivalent [34]. Finally, the following equations in Table 6 were used to determine energy indicators such as energy use efficiency/energy yield, energy productivity, specific energy, mechanization index, net energy, and agrochemical energy ratio [40, 57, 74].

2.3.2 Modeling of Life Cycle GHG Emission Analysis

According to ISO 14040 LCA approach, LCA addresses quantitative assessment methods for assessing the environmental aspects of a product or service throughout its entire life cycle stages, which defines four conceptual steps, including goal and scope definition, life cycle inventory (LCI), life cycle impact assessment, and interpreting results [25, 68]. The LCA approach was applied to estimate the global warming impact (GWI) in terms of total GHG emissions of agricultural products, as reported by the recent literature [2, 22, 68]. Then, the net life cycle GHG emissions were determined after subtracting SOC accumulation from overall GHG emissions.

Goal Setting and Scope Definition

The study aims to conduct the GHG emissions assessment of wheat production, distribution of the wheat grain to the storage and end-of-life treatment. A functional unit is a unit of measurement used to calculate the function of a system and its environmental implications, where the functional unit, are commonly defined by the weight or volume of a product, used for this study as one kilogram per ton of wheat grains produced by a CA-based wheat production process (Fig. 1). For this, the LCA system boundary for wheat farming was employed from cradle to farmgate, which has both off-farm and on-farm stages. The system boundary encompasses all inputs used in the production process and releases of all sorts of emissions. This study intended to identify emission impact categories and compare the influence of input energy on environmental sustainability. The inputs were categorized as system inputs, and the product produced along with the release of emission were categorized as system outputs. The detailed agricultural modeling of wheat production farming was not included in the scope of the study; a readily available dataset was applied to include the contribution of wheat farming in the overall process.

Table 4 Energy input-outputs usage in wheat farming

Operations	Input	Unit	Strip tillage (ha)	Minimum tillage (ha)	Conventional tillage (ha)
Land preparation	Power Tiller (PT)	h	–	–	24.7
	PTOS	h	11.2	11.2	–
	Diesel Fuel	l	13.5	13.5	24.7
	Human Labor	h	16.5	16.5	28.0
Planting	Seed	kg	119.8	119.8	164.67
	Human Labor	h	0.0	0.0	16.5
Fertilization	Nitrogen (N)	kg	90.0	90.0	90.0
	Phosphate (P_2O_5)	kg	80.0	80.0	80.0
	Potassium (K_2O)	kg	60.0	60.0	60.0
	Sulfur (S)	kg	85.0	85.0	85.0
	Zinc (Z)	kg	10	10	10
	Boron (B)	kg	5.2	5.2	5.2
	Manure	kg	3750	3750	3750
	Human Labor		16.5	16.5	16.5
Irrigation	Electricity	Kwh	34	34	34
	Irrigation water	m3	2625.8	2625.8	2625.8
	Human Labor	h	49.4	49.4	49.4
Weeding	Human Labor	h	98.8	148.2	115.3
Spraying	herbicide	kg	1.6	1.6	1.6
	Human Labor	h	16.5	24.7	24.7
Harvesting	Combine Harvester	h	2.5	2.5	2.5
	Diesel Fuel	L	17.5	17.5	17.5
	Human labor	h	9.9	9.9	9.9
	Wheat Grains	kg	4014.9	4082.6	3820.5
	Wheat Straws	kg	5356.93	5486.8	5018.3

Table 5 Energy equivalents for different inputs and outputs energy in wheat farming

Operations	Parameters	Unit	Energy equivalent (MJ/unit)	Reference no
A. Inputs				
Tillage	Machinery	h	62.7	[80]
	Human labor	h	1.96	[47]
	Diesel fuel	l	56.31	[23, 76]
Seeding	Seed	kg	14.7	[41]
Fertilization	(a) Nitrogen (N)	kg	64.14	[59]
	(b) Phosphate (P_2O_5)	kg	12.44	[59]
	(c) Potassium (K_2O)	kg	11.15	[59]
	d) Zinc (Z)	kg	5.0	[41]
	e) Sulfur (S)	kg	1.12	[41]
Herbicide	Herbicides	kg	101.2	[51]
	Electricity	kWh	3.6	[41]
	Water for irrigation	m^3	1.02	[41]
B. Output				
	Wheat grain	kg	14.7	[41, 80]
	Wheat biomass	kg	18.6	[41, 42, 55]

Table 6 Equations followed to calculate the energy indicators

Energy use efficiency = Wheat grains yield($kgha^{-1}$)/Input Energy($MJha^{-1}$)	(2)
Energy productivity = Output energy($MJha^{-1}$)/Input Energy($MJha^{-1}$)	(3)
Specific energy = Input Energy($MJha^{-1}$)/Wheat yield($kgha^{-1}$)	(4)
Net energy = Outputenergy($MJha^{-1}$) − Input Energy($MJha^{-1}$)	(5)
Agrochemical energy ratio(%) = Agrochemicals input Energy($MJha^{-1}$)/Input energy($MJha^{-1}$)	(6)

Life Cycle Inventory (LCI)

In these steps, a systematic approach to LCI is a technique for collecting, analyzing, and assessing total inputs and outputs (including emissions into the air, water, soil, and waste processes) flow of each production process while taking into account a system boundary (Fig. 2) [48]. In this stage, the amount of inputs flow, including material manufacturing and transport to the farm gate, and outputs flow for one functional unit of wheat grains to establish a complete LCI by applying a mass balance approach with LCA system boundary (Table 4) [2]. Here, the GHG emissions consisting of CO_2, CH_4, and N_2O were the negative outputs, and SOC accumulation was the positive outputs considering the LCA system boundary of this study. The GHG emissions observed within the LCI system boundary is mainly from off-firm and on-firm activities.

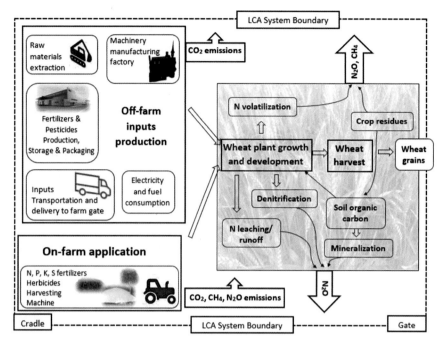

Fig. 2 LCA system boundary including off-firm, on-firm, growth, and development stages and after harvesting activities to analyze environmental aspects of CA wheat farming

Off-farm GHG emissions: Activities of agricultural inputs production and delivery up to farmgate related to GHG emissions were estimated, known as off-farm GHG emissions. The input and output database based on the LCA study for wheat farming was used to determine the farm machinery manufacturing responsible for indirect emissions [66]. The input and output database based on the LCA study for wheat farming was used to determine the farm machinery manufacturing responsible for indirect emissions [66]. Farm machinery's emission factor (EF) was determined by multiplying the machinery production cost for each functional unit, validating the amount provided in the reference [2]. The EFs of chemicals and herbicides used for wheat farming were sourced from the Bangladesh studies [2, 15], and the EFs of imported inputs to Bangladesh were derived from the studies, including fertilizer materials [2], as these data are defined for the local condition in Bangladesh. The GHG emissions data associated with materials transportation for one functional unit of wheat production were used from available databases of EFs of Road Transport [50].

The transportation vehicle for road transport was used in medium-sized trucks of 7 tons and for sea transport was used a medium-size cargo carrier. The GHG emissions associated with the activities of inputs delivery from storage factory to field gate is expressed as ton-kilometers (tkm) for road transport and ton nautical miles (tnm) for sea transport. The inputs weight is multiplied by the distance between the wheat study field and storage factory for determining tkm and tnm [2, 43, 50, 78].

On-farm GHG emissions: On-farm GHG emissions was related to the on-farm activities for establishing wheat farming. The first emi ssion associated with diesel fuel use by farm machinery is for the preparation of land. Agricultural machinery, such as BARI Seeder, power tiller, and combine harvester, were used for land preparation, seed sowing and harvesting of different treatments of wheat farming. The GHG emissions was assessed for the application of chemicals and weeding in this stage. The fuel used was calculated in standard machinery usage terms (9.35 lha^{-1} for PTOS & BARI Seeder, 25.5 lha^{-1} for power tiller and 1.82–2.11 lton^{-1} for combine harvester). The direct GHG emissions from the soil related to CO_2, CH_4, and N_2O were estimated using the database study owing in FEAT tool. The N_2O emitted directly described in IPCC [65] and also emitted indirectly by ammonia volatilization, runoff of nitrate, crop residue in aboveground and belowground were examined from the database study owing in FEAT software.

Impact Assessment

By multiplying the emission factors (EF) with the associated inputs and outputs in Table 4, the total GHG emissions (CO_2, CH_4, and N_2O) for producing one functional unit of wheat production were determined. This total GHG emissions were converted to CO_2eq GHG emissions using the EF for the respective type of emission or global warming potential (GWP) by applying an IPCC method [65]. The CO_{2eq} GHG emissions for producing each functional unit of wheat production was determined based on GWP value of 100 years accordingly IPCC 2013, where the EFs of 25 and 298 were used for CH_4 and N_2O, respectively [65]. Finally, the following equation (7) was applied to determine net life cycle GHG emissions:

$$NGHG = TGHG - SOCA \qquad (7)$$

where

NGHG represents the net life cycle GHG emissions (tonCO$_2$eq ha^{-1})
TGHG represents the total GHG emissions (tonCO$_2$eq ha^{-1}) and
SOCA represents the SOC accumulation in unit of land (kgCO$_2$eq ha^{-1}).

Interpreting Results

The last step of the life cycle GHG emissions methodology is the interpretation of results. The investigated results were interpreted and a suitable model to reduce life cycle GHG emissions was proposed, which is shown in Sect. 3.

2.4 Data Analysis

The different tools and software were employed to analyze the data, which was collected from the experimental field in this study. The STAR statistical analysis software developed by IRRI was employed to analyze yield parameters, where LSD value was assumed to compare means considering a 5% significance level. The FEAT tool was employed to estimate energy use analysis and life cycle GHG emission to determine the energy and carbon footprint.

3 Results and Discussion

3.1 Energy Use Analysis

Table 7 shows the energy use analysis of the different inputs and outputs energy to different tillage practices (ST, MT, and CT). The results show that inputs energy consumption during the lifecycle of wheat farming for ST, MT, and CT practices were 16,268.77, 16,233.26, and 18,068.80 MJ ha^{-1}, respectively. The researcher reported in the previous studies that total energy input has been recorded to be 18,680.8 MJ ha^{-1}, 14,358 MJ ha^{-1}, and 16,000.36 MJ ha^{-1} for wheat farming [16, 49, 59]. In this study, energy consumption for ST and MT practices were relatively low compared to the CT practice because energy inputs of ST practice for land preparation by machinery and diesel fuel used to different management operations were comparatively low. Moreover, wheat grains yield was found to be 4015, 4083, and 3821 kg for ST, MT, and CT practices, accordingly, while the energy output was calculated to be 158,657.40, 162,069.55, and 149,501.58 MJ ha^{-1}, respectively. It is seen that CA-based tillage practices were adopted to reduce the requirement of diesel fuel energy aiming at improving energy use efficiency (EUE). However, CA tillage practices were used to reduce soil microbial activities to decrease CO_2 emissions further, results in increasing agricultural sustainability. For this, conservation tillage practices can be considered an energy-efficient technology and positively minimize energy consumption. Among the energy input of conservation tillage practices and energy output yield, it is shown that MT practice consumes less energy inputs which can be considered one of the sustainable indicators for achieving environmental sustainability.

The table also shows that the most significant part of inputs energy was consumed by synthetic fertilizer of 46.6, 46.71, and 46.96% for ST, MT, and CT, respectively, where N fertilizer share is for 35.48, 35.56, and 31.95%, respectively. The nitrogen fertilizer for all the treatments is the largest energy consumer among the input's energy. Alongside this fertilizer energy, the latter most immense input energy in wheat farming was irrigation water, with 16.46, 16.5, and 14.82% for ZT, MT, and CT practices, respectively. Electricity, human labor, herbicide, and machinery were observed as the minor energy inputs.

The researcher observed that the most significant part of inputs energy consumption among the chemical energy is the N fertilizer which was around 40% of inputs energy [28, 33, 64] because it is an essential element as a moderator in plant growth and biological development and increasing crop yield if applied at the right time. Besides this, N fertilizer has an adverse effect linearly on growth and crop productivity if this fertilizer is not applied in a proper way and right time. This study shows that N fertilizer was the primary input energy consumption in conventional farming, distinguishing between conventional and sustainable agricultural systems. Chemical fertilizer use grew in CT practice, resulting in a higher share of non-renewable energy use, and this matter had seen entirely negative from the philosophy of agroecology consequences adverse impact of CT practice on agricultural sustainability. Therefore, the alternative measured had taken for the renewal or lesser use of N fertilizer. This study shows that renewal was done using cowdung with the use of less N fertilizer (90 kg) than recommended (120 kg) (Table 6), thus input energy consumption was also less with the use of this fertilizers. The energy value of cowdung is equal to 1115 MJ and its equivalent energy of N fertilizer is found from only 17.8 kg, whereas the N fertilizer was applied 90.0 kg per year (avg.). Moreover, the results show that wheat grain productivity was higher in ST and MT practices compared to the CT practice. For this, sustainable fertilizer management had a positive effect to supply more nutrition in soil than chemical fertilizer, getting a more environmental sustainability. As a result, it's worth noting that using manure with biofertilizer in wheat farming can be a viable alternative to reducing energy consumption and developing a sustainable production system that boosts productivity and finally ensures agricultural sustainability. Moreover, CA-based management practices, like composts and stable crop residues management, may increase SOC content and soil health, which will reduce the chemical fertilizer energy demand.

3.2 Energy Indicators in Wheat Farming

Energy indicators in wheat farming were determined by applying the equations in Table 6 for ST, RT, and CT practices (Table 8). The results show that energy use efficiency (EUE) in wheat farming was 9.75, 9.98, and 8.27 for ST, RT, and CT practices, respectively. The EUE is shown in higher in comparison to the stated 2.8 for wheat production systems in Turkey [16], 1.44 in south Panjab in Pakistan [34], 2.9 to 5.2 for wheat production in India [63], and 2.3 in Bangladesh [59]. Comparatively, Lower EUE was observed in CT systems in wheat production due to intake of higher energy inputs than the ST and MT practices, whereas the higher EUE was shown for the MT practice. Energy productivity in wheat farming was calculated as 0.247, 0.251, and 0.211 kg MJ^{-1} for ST, MT, and CT practices, respectively. This corresponds to 247, 251, and 211 g of wheat grain produced per MJ of input energy use for ST, RT, and CT practices, respectively. Previous studies reported that energy productivity in wheat farming was 0.11 kg MJ^{-1} for high inputs and 0.14 kg MJ^{-1} for low inputs usage [75]. Analysis showed that energy productivity is comparatively higher than previously reported in all the studies in wheat farming. The net energy balance in wheat farming for ST, RT, and CT practices was calculated to be 142,388.63, 145,836.29, and 131,432.77 MJ ha^{-1}, respectively. Energy productivity and net energy per ha in Bangladesh, according to Sanzidur and Kamrul, were 0.2 and 20,595.9 MJ ha^{-1}, respectively [59]. Due to higher yields (grain and straw) with comparatively lower inputs, net energy was found higher in this study than in the previous study for wheat farming. Additionally, energy indices of agrochemical energy, which was found to be 55, 53, and 49% for ST, RT, and CT practices, respectively. This indicator implies that comparatively less agrochemical energy is applied in the ST practice resulted in showing a higher percentage. The modeling of these indicators can be applied in wheat farming systems to use energy efficiently for achieving higher yield, productivity, and sustainability.

Wheat farming in Bangladesh is completely energy efficient in terms of estimated net energy balance and energy ratio, although these indicators can increase further by less inorganic fertilizer. Thus, the sustainability test had been successfully evaluated in terms of energy balance in wheat farming, which indicates that this farming is more energy-efficient and sustainable.

3.3   Assessment of Life Cycle GHG Emission and Carbon Footprint

Figures 3 and 4 show the total life cycle GHG emissions for wheat farming inputs when the LCA system boundary is taken into account. Total GHG emissions for different tillage practices linked with fixed crop residue retention were calculated using a single global warming potential (GWP) allocation in wheat farming. The results show that total GHG emissions for ST, MT, and CT practices were calculated to be 1.987, 1.992, and 2.028 tonCO$_2$eq ha^{-1}, respectively; whereas 0.495, 0.488, and 0.531 tonCO$_2$eq ton^{-1} of wheat grains, respectively. Nitrous oxide (N$_2$O) accounted for the greatest share of total GHG emissions, with about 562 kgCO$_2$eq ha^{-1} of wheat grain accounting for 28.28, 28.21, and 27.71% of total GHG emissions for ST, MT, and CT practices, respectively.

Furthermore, N fertilizer was found to be the second largest contributor in ST, MT, and CT practices, accounting for 24.98, 24.92, and 24.47%, respectively, followed by N$_2$O from aboveground crop residues, nitrous oxide leaching/runoff, diesel fuel, Phosphate, N$_2$O from belowground crop residues, N$_2$O from manure, N$_2$O volatilization, seed, potash, herbicide, input transportation, and electricity. N fertilizer and N$_2$O are the primary consumers of GHG emissions, with some of the nitrogen being dissolved aboveground in the form of volatilization and the rest being applied dissolved into the soil, where it is naturally transformed to N$_2$O by soil microbes. Previous research revealed that wheat cultivation in Pakistan produced 1.118 ton CO$_2$eq ha^{-1} of GHG emissions [34]. According to other research, China's GHG emissions were 2.75 ton CO$_2$eq ton^{-1} of wheat production [78]. According to this

Table 7 The amount of input and output energy utilized in different tillage processes, as well as their percentage share in wheat farming

Input	ST (MJ/ha)	Percentage (%)	MT (MJ/ha)	Percentage (%)	CT (MJ/ha)	Percentage (%)
Human labor	229.15	1.41	193.65	1.19	284.02	1.59
Machinery	860.70	5.21	860.70	5.3	1498.95	8.3
Diesel Fuel	1744.07	11.72	1744.07	10.74	2190.83	12.12
Seed	1760.44	10.82	1760.44	10.84	2420.60	13.4
(a) Nitrogen	5772.60	35.48	5772.60	35.56	5772.60	31.95
(b) Phosphate (P_2O_5)	995.20	6.12	995.20	6.13	995.20	5.51
(c) Potassium (K_2O)	669.00	4.11	669.00	4.12	669.00	3.7
(d) S	95.20	0.59	95.20	0.59	95.20	0.53
(e) Z	50.00	0.31	50.00	0.31	50.00	0.28
(g) Manure	1125	6.92	1125	6.93	1125.0	6.23
Electricity	122.40	0.75	122.40	0.75	122.40	0.68
Irrigation water	2678.36	16.46	2678.36	16.5	2678.36	14.82
Herbicide	166.64	1.02	166.64	1.03	166.64	0.92
Total input energy	16,268.77	100.00	16,233.26	100.00	18,068.80	100.00
Grain	59,018.55	37.20	60,014.25	37.03	56,162.0	37.57
Straw	99,638.85	62.80	102,055.31	62.97	93,339.5	62.43
Total output energy	158,657.40	100.00	162,069.55	100	149,501.58	100

Note ST = Strip tillage practice, MT = Minimum tillage practice, CT = COnvertional tillage practice

study, the use of less nitrogen fertilizer results in a favorable reduction in GHG emissions. Furthermore, as demonstrated by the experiment, stable crop residue retention improved soil structure and increased soil organic carbon (SOC) content by pulverizing it into the soil using conservation tillage practices (CTP), which helped fix atmospheric CO_2 into the soil, resulting in a reduction in total GHG emissions. Soil fertility and environmental quality were significantly improved by implementing CTPs, such as ST and MT practices, and balanced fertilizer management with stable crop residue retention, in order to reduce GHG emissions.

The relative contributions of different tillage treatments at the off-farm stage were nearly similar to GHG emissions for all treatments. The off-farm activities emitted

Table 8 Results of energy indicators in wheat cultivation using ST, MT, and CT practices

a	Unit	ST	(%)	MT	(%)	CT	(%)
Energy use efficiency/Net energy ratio	kg MJ^{-1}	9.75		9.98		8.27	
Energy productivity	kg MJ^{-1}	0.247		0.251		0.211	
Specific energy	MJ kg^{-1}	3.91		3.84		3.84	
Net energy	MJ ha^{-1}	142,388.63		145,836.29		131,432.77	
Agrochemical energy ratio	(%)	0.55		0.53		0.49	
Mechanization Index	(%)	17.2		17.3		15.5	
Direct energy *	MJ ha^{-1}	4651.59	28.59	4616.08	28.44	5153.21	28.52
Indirect energy **	MJ ha^{-1}	11,617.18	71.41	11,617.18	71.56	12,915.59	71.48
Renewable energy ***	MJ ha^{-1}	4667.95	28.69	4632.45	28.54	5382.98	29.79
Non-renewable energy ****	MJ ha^{-1}	11,600.82	71.31	11,600.82	71.46	12,685.83	70.21

Note *Direct energy, **Indirect energy, ***Renewable energy, ****Non-renewable energy

GHGs were approximately 37.97, 37.88, and 38.3% of total GHG emissions for ST, MT, and CT practices, respectively (Fig. 5). A slight increase in emission was observed only for CT practice because wheat seeds and fuels used in this treatment are slightly more. Emission associated with different tillage practices and fixed crop residue retention were not significantly different in off-farm treatment tillage practices.

Analysis in the on-farm stage confirmed that there had not been any CH$_4$ emissions, having the highest GHG emissions from the CT practice than the ST and MT practices, whereas GHG emissions from ST and MT practices were showed similar trends (Fig. 5). The only variation in GHG emissions is from the fuel and biological seed emissions. On-farm GHG emissions can be reduced by reduce use of N fertilizer and increase use of organic fertilizer. Overall, off-farm GHG emissions of wheat farming were much lower than emitted GHG emissions during the on-farm stage because of adopting CA practices.

The GHG intensity or carbon footprint (CF) of wheat grains was determined to be 0.013, 0.012, and 0.014 kgCO$_2$eq MJ^{-1} for ST, MT, and CT practices, respectively. In previous investigations, CF was found to be 0.027 for wheat in the United States [15]. Due to reduce use of N fertilizer and conservation tillage measures, CF was lower in this study. CA practices limit the amount of fuel used in tillage operations,

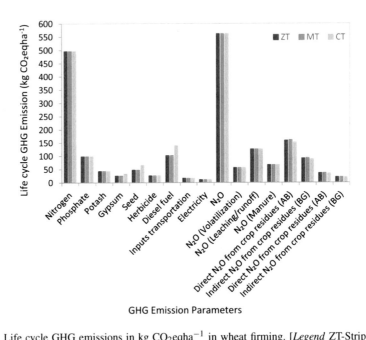

Fig. 3 Life cycle GHG emissions in kg CO_2eqha^{-1} in wheat firming. [*Legend* ZT-Strip tillage; MT-Minimum tillage; CT-Conventional tillage practice]

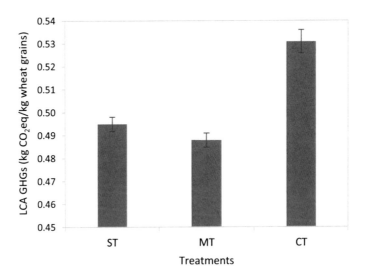

Fig. 4 Total life cycle GHG emissions for one ton of wheat produced per hectare in wheat farming ($p<0.05$)

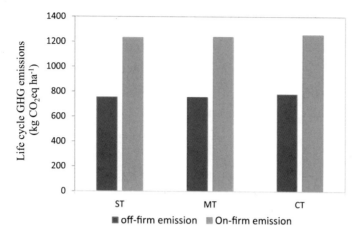

Fig. 5 Life cycle greenhouse gas emissions for one functional unit for off-firm and on-firm activities

resulting in reduced CF and environmental sustainability. The combined effect of improved nitrogen use efficiency, reduced life cycle GHG emissions, and lower CF has been the key result of lowering GHG intensity in wheat farming (Fig. 5).

The net life cycle GHG emissions must be calculated considering SOC accumulation during crop farming, because total life cycle GHG emissions must be accounted for off-farm and on-farm GHG emissions. In agricultural farming, the effects of SOC accumulation on GHG emissions must be recognized; otherwise, life cycle GHG emission results would be misleading. The accumulation of SOC linked with CA-based management practices might take several years to optimize. As a result, when soil characteristics have reached an equilibrium state with the new degree of SOC, the net life cycle GHG emissions of wheat farming using CA-based establishment practices must be estimated and give inceptive results.

3.3.1 Soil Organic Carbon (SOC) Accumulation

The topsoil tends to have the highest concentration of soil organic carbon (SOC) percentage. To determine how much SOC buildup by adopting conservation tillage practices, topsoil up to 20 cm was linked with measurement. Table 9 shows how the SOC accumulation was translated to kilograms of CO_2 equivalent per hectare. For ST, MT, and CT practices, SOC accumulation was 18.94, 14.93, and 5.66 $kgCO_2eq\ ha^{-1}$, respectively. After converting this equivalent SOC accumulation to equivalent CO_2, net GHG emissions for ST, MT, and CT practices were 1.968, 1.977, and 2.023 $kgCO_2eq\ ha^{-1}$, respectively. Although the quantity of SOC per hectare in terms of a kilogram of CO_2 equivalent is modest, this amount of SOC accumulation is only for two years of wheat farming. According to the research, crop residues retention in soil, low soil disturbance, reduced CO_2 emissions, and crop rotation with legume crops can all help to improve SOC content [3, 8]. The study result

Table 9 SOC accumulation and their GHG emissions in wheat farming

Treatment	Total GHG emission (kgCO$_2$eq/ha)	SOC (t/ha)	SOC accumulation (t/ha)	SOC (kgCO$_2$eq/ha)	Net GHG Emissions (kgCO$_2$eq/ha)	Net GHG Emissions (tonCO$_2$eq/ha)
Initial soil		16.240				
ST	1987.22	21.291	5.051	18.94	1968.28	1.968
MT	1992.01	20.221	3.981	14.93	1977.08	1.977
CT	2028.28	17.750	1.510	5.66	2022.62	2.023

implied that CA-based tillage practices and stable crop residue retention enhance SOC in the topsoil.

3.4 Net Life Cycle GHG emission

Net life cycle GHG emissions for different tillage practices were evaluated based on overall GHG emissions for the production of 1 ton of wheat grains after accounting for SOC accumulation. According to the findings, SOC accumulation has a significant impact on lowering net life cycle GHG emissions. The ST practice was also shown to be the best practice for reducing GHG emissions since the highest amount of soil was covered and untilled following during the tillage operation. However, SOC content might reduce GHG emissions in the CT practice to some extent, as heavy tillage in the soil causes intensive soil disturbance and a high level of microbial activity, resulting in less carbon remaining in this practice.

4 Conclusion

The input–output energy balance in wheat farming is a systematic approach to utilize available inputs effectively for increasing overall farming efficiency and environmental sustainability. This study was conducted within the LCA system boundary to evaluate the energy balance of wheat farming using different conservation tillage practices with a stable crop residue retention used in Bangladesh. The analysis of energy use was performed using input and output energy to determine energy use efficiency, net energy, energy productivity, and agrochemical energy ratio for ST, MT, and CT practices, respectively. It is revealed that a lower amount of input energy was related to ST and MT practices, whereas a higher amount was associated with CT practices in wheat farming. Energy use efficiency (EUE) and productivity in wheat farming showed higher value with MT practice (9.98 and 0.251 kg MJ^{-1}) than CT practice (8.27 and 0.211 kg MJ^{-1}). Hence, EUE in wheat farming for different tillage practices kept to the MT > ST > CT sequence. A FEAT tool with LCA system boundary was employed for estimating life cycle GHG emissions ($tonCO_2eq\ ha^{-1}$) and carbon footprint of wheat farming considering the whole production process, including cradle to farmgate activities. The estimated life cycle GHG emission in wheat farming was 1.968, 1.977, and 2.023 $tonCO_2eq\ ha^{-1}$ for ZT, MT, and CT, respectively. Nitrous oxide (N_2O) direct emission, N fertilizer, N_2O from above-ground crop residues and fuel for machinery operation had the significant suppliers in total GHG emissions. Additionally, the GHG intensity or carbon footprint was 0.013, 0.012, and 0.014 $kgCO_2eq\ MJ^{-1}$ of output energy from wheat farming for ST, MT, and CT practices, respectively, which means the farming system is environmentally more sustainable.

The highest emissions from wheat farming inputs were N fertilizer (41.1% for ST and MT practices, 37.43% for CT practices), followed by irrigation water, diesel, and seed. To address this problem, biofertilizers like manure or cow dung might be suggested to improve soil nutrition and fertility, potentially increasing SOC accumulation and reducing net life cycle GHG emissions. Adopting the conservation agriculture (CA) practice for reducing GHG emissions is also an effective measure. CA-based conservation tillage practices promote SOC accumulation by reducing input energy consumption by taking less fuel and retaining stable crop residue at a certain level. Both these plays a vital role in reducing GHG emissions and achieving long-term environmental sustainability.

Acknowledgements The authors would like to acknowledge the Bangladesh Agricultural Research Institute (BARI) authority for financial support and the Regional Laboratory of the Soil Resources Development Insitute (SRDI), Jamalpur for laboratory supports.

References

1. Adhya TK, Mishra SR, Rath AK, Bharati K, Mohanty SR, Ramakrishnan B, Rao VR, Sethunathan N (2000) Methane efflux from rice-based cropping systems under humid tropical conditions of eastern India. Agric Ecosyst Environ. https://doi.org/10.1016/S0167-8809(99)00144-9

2. Alam MK, Bell RW, Biswas WK (2019) Decreasing the carbon footprint of an intensive rice-based cropping system using conservation agriculture on the Eastern Gangetic Plains. J Clean Prod 218:259–272. https://doi.org/10.1016/j.jclepro.2019.01.328

3. Alam MK, Bell RW, Haque ME, Kader MA (2018) Minimal soil disturbance and increased residue retention increase soil carbon in rice-based cropping systems on the Eastern Gangetic Plain. Soil Tillage Res 138:28–41. https://doi.org/10.1016/j.still.2018.05.009

4. Alam MK, Biswas WK, Bell RW (2016) Greenhouse gas implications of novel and conventional rice production technologies in the Eastern-Gangetic plains. J Clean Prod 112:3977–3987. https://doi.org/10.1016/j.jclepro.2015.09.071

5. Alam MK, Islam MM, Salahin N, Hasanuzzaman M (2014) Effect of tillage practices on soil properties and crop productivity in wheat-mungbean-rice cropping system under subtropical climatic conditions. Sci World J 7:1–17. https://doi.org/10.1155/2014/437283

6. Arunrat N, Pumijumnong N, Sereenonchai S, Chareonwong U (2020) Factors controlling soil organic carbon sequestration of highland agricultural areas in the Mae Chaem Basin, northern Thailand. Agronomy 10:305. https://doi.org/10.3390/agronomy10020305

7. Azad AK, Miaruddin M, Wohab MA, Sheikh MHR, Nag BL, Rahman MHH (2020) KRISHI PROJUKTI HATBOI (Handbook on Agro-Technology), 9th edn. Bangladesh Agricultural Research Institute, Gazipur-1701, Bangladesh

8. Baldock JA (2007) Composition and cycling of organic carbon in soil. In: Nutrient cycling in terrestrial ecosystems, pp 1–35. https://doi.org/10.1007/978-3-540-68027-7_1

9. Banerjee A, Jhariya MK, Raj A, Yadav DK, Khan N, Meena RS (2021) Energy and climate footprint towards the environmental sustainability. In: Agroecological footprints management for sustainable food system. https://doi.org/10.1007/978-981-15-9496-0_14

10. Bangladesh Bureau of Statistics (BBS) (2018) Yearbook of agricultural statistics-2017

11. Bell RW, Enamul Haque M, Jahiruddin M, Moshiur Rahman M, Begum M, Monayem Miah MA, Ariful Islam M, Anwar Hossen M, Salahin N, Zahan T, Hossain MM, Alam MK, Mahmud MNH (2019) Conservation agriculture for rice-based intensive cropping by smallholders in the eastern gangetic plain. Agric 9:1–17. https://doi.org/10.3390/agriculture9010005

12. Bhatia A, Sasmal S, Jain N, Pathak H, Kumar R, Singh A (2010) Mitigating nitrous oxide emission from soil under conventional and no-tillage in wheat using nitrification inhibitors. Agric Ecosyst Environ. https://doi.org/10.1016/j.agee.2010.01.004

13. Black CA (1950) Method of soil analysis Part I and II. Am Soc Agron Inc 770

14. Burt R (2014) Soil survey field and laboratory methods manual soil survey investigations report No. 51. United States Dep. Agric. Nat. Resour. Conserv. Serv. version 2

15. Camargo GGT, Ryan MR, Richard TL (2013) Energy use and greenhouse gas emissions from crop production using the farm energy analysis tool. Bioscience 63:263–273. https://doi.org/10.1525/bio.2013.63.4.6

16. Canakci M, Topakci M, Akinci I, Ozmerzi A (2005) Energy use pattern of some field crops and vegetable production: Case study for Antalya Region, Turkey. Energy Convers Manag 46:655–666. https://doi.org/10.1016/j.enconman.2004.04.008

17. Chakrabarti B, Pramanik P, Mina U, Sharma DK, Mittal R (2014) Impact of conservation agricultural practices on soil physic-chemical properties. Int J Agric Sci 5:55–59

18. Chatterjee S, Ghosh S, Pal P (2020) Soil carbon restoration through conservation agriculture. In: Natural resources management and biological sciences. https://doi.org/10.5772/intechopen. 93006

19. Chowdhury MAH, Hassan MS (2013) Agricultural technology hand book. Bangladesh Agricultural Research Council, Farmgate, Dhaka-1215

20. Clark A (2007) Managing cover crops in conservation tillage systems, 3rd edn. In: Sustainable Agriculture Research and Education (SARE) program. National Institute of Food and Agriculture, U.S. Department of Agriculture, Maryland, USA

21. Dachraoui M, Sombrero A (2020) Effect of tillage systems and different rates of nitrogen fertilisation on the carbon footprint of irrigated maize in a semiarid area of Castile and Leon, Spain. Soil Tillage Res 196:104472. https://doi.org/10.1016/j.still.2019.104472

22. Denham FC, Biswas WK, Solah VA, Howieson JR (2016) Greenhouse gas emissions from a Western Australian finfish supply chain. J Clean Prod 112:2079–2087. https://doi.org/10.1016/j.jclepro.2014.11.080

23. Erdal G, Esengün K, Erdal H, Gündüz O (2007) Energy use and economical analysis of sugar beet production in Tokat province of Turkey. Energy 32:35–41. https://doi.org/10.1016/j.energy.2006.01.007

24. FAOSTAT (2021) Emiss Shares, Foog Agric. Organ. United Nations. http://www.fao.org/faostat/en/#data/EM/metadata. Accessed 8 Jan 21

25. Finkbeiner M, Inaba A, Tan RBH, Christiansen K, Klüppel HJ (2006) The new international standards for life cycle assessment: ISO 14040 and ISO 14044. Int J Life Cycle Assess. https://doi.org/10.1065/lca2006.02.002

26. Forte A, Fiorentino N, Fagnano M, Fierro A (2017) Mitigation impact of minimum tillage on CO_2 and N_2O emissions from a Mediterranean maize cropped soil under low-water input management. Soil Tillage Res 166:167–178. https://doi.org/10.1016/j.still.2016.09.014

27. Furtak K, Gajda AM (2018) Activity and variety of soil microorganisms depending on the diversity of the soil tillage system. In: Sustainability of agroecosystems, p 45. https://doi.org/10.5772/intechopen.72966

28. Giampietro M, Cerretelli G, Pimentel D (1992) Energy analysis of agricultural ecosystem management: human return and sustainability. Agric Ecosyst Environ 38:219–244. https://doi.org/10.1016/0167-8809(92)90146-3

29. Gómez-Limón JA, Sanchez-Fernandez G (2010) Empirical evaluation of agricultural sustainability using composite indicators. Ecol Econ 69:1062–1075. https://doi.org/10.1016/j.ecolecon.2009.11.027

30. He L, Zhang A, Wang X, Li J, Hussain Q (2019) Effects of different tillage practices on the carbon footprint of wheat and maize production in the Loess Plateau of China. J Clean Prod 234:297–305. https://doi.org/10.1016/j.jclepro.2019.06.161

31. Hillier J, Hawes C, Squire G, Hilton A, Wale S, Smith P (2009) The carbon footprints of food crop production. Int J Agric Sustain 7:107–118. https://doi.org/10.3763/ijas.2009.0419

32. Hossain MI, Sarker M, Haque MA (2015) Status of conservation agriculture based tillage technology for crop production in Bangladesh. Bangladesh J Agric Res 40:235–248. https://doi.org/10.3329/bjar.v40i2.24561

33. Hosseinpanahi F, Kafi M (2012) Assess the energy budget in farm production and productivity of potato (*Solanum tuberosum* L.) in Kurdistan, case study: plain Dehgolan. J Agroecol 4:159–169

34. Imran M, Özçatalbaş O, Bashir MK (2020) Estimation of energy efficiency and greenhouse gas emission of cotton crop in South Punjab, Pakistan. J Saudi Soc Agric Sci 19:216–224. https://doi.org/10.1016/j.jssas.2018.09.007

35. ISO (2016) Environmental management—life cycle assessment—requirements and guidelines ISO 14044:2006. Int Organ Stand. https://www.iso.org/standard/38498.html. Accessed 23 Oct 21

36. Jackson ML (1959) Soil chemical analysis. J Agric Food Chem 7:138. https://doi.org/10.1021/jf60096a605

37. Jain N, Dubey R, Dubey DS, Singh J, Khanna M, Pathak H, Bhatia A (2014) Mitigation of greenhouse gas emission with system of rice intensification in the Indo-Gangetic Plains. Paddy Water Environ. https://doi.org/10.1007/s10333-013-0390-2

38. Jat ML, Saharawat Y, Gupta R (2011) Conservation agriculture in cereal systems of south Asia: nutrient management perspectives. J Agric Sci 24:100–105

39. Jiban S, Subash S, Prasad TK, Amit C, Manoj K, Subina T (2020) Conservation agriculture as an approach towards sustainable crop production : a review. Farming Manag 5:7–15. https://doi.org/10.31830/2456-8724.2020.002

40. Khan S, Khan MA, Hanjra MA, Mu J (2009) Pathways to reduce the environmental footprints of water and energy inputs in food production. Food Policy 34:141–149. https://doi.org/10.1016/j.foodpol.2008.11.002

41. Khosruzzaman S, Asgar MA, Rahman KR, Akbar S (2010) Energy intensity and productivity in relation to agriculture-bangladesh perspective. J Bangladesh Acad Sci 34:59–70. https://doi.org/10.3329/jbas.v34i1.5492

42. Kurkela E, Moilanen A, Nieminen M (1999) CFB gasification of biomass residues for co-combustion in large utility boilers: studies on ash control and gas cleaning. In: Ower production from biomass III: gasification and pyrolysis R&D&D for Industry. VTT Technical Research Centre of Finland. VTT Symposium, No. 192, Espoo, Finland, pp 213–228

43. Lal R (2004) Carbon emission from farm operations. Environ Int. https://doi.org/10.1016/j.envint.2004.03.005

44. Liu C, Wang K, Meng S, Zheng X, Zhou Z, Han S, Chen D, Yang Z (2011) Effects of irrigation, fertilization and crop straw management on nitrous oxide and nitric oxide emissions from a wheat-maize rotation field in northern China. Agric Ecosyst Environ. https://doi.org/10.1016/j.agee.2010.12.009

45. Liu J, Wang H, Rahman S, Sriboonchitta S (2021) Energy efficiency, energy conservation and determinants in the agricultural sector in emerging economies. Agric 11:773. https://doi.org/10.3390/agriculture11080773

46. Mittal R, Chakrabarti B, Jindal T, Tripathi A, Mina U, Dhupper R, Chakraborty D, Jatav RS, Harit RC (2018) Carbon footprint is an indicator of sustainability in rice-wheat cropping system: a review. Chem Sci Rev Lett 7:774–784

47. Mohammadshirazi A, Akram A, Rafiee S, Mousavi Avval SH, Bagheri Kalhor E (2012) An analysis of energy use and relation between energy inputs and yield in tangerine production. Renew Sustain Energy Rev 16:4515–4521. https://doi.org/10.1016/j.rser.2012.04.047

48. Naderi SA, Dehkordi AL, Taki M (2019) Energy and environmental evaluation of greenhouse bell pepper production with life cycle assessment approach. Environ Sustain Indic 3–4:100011. https://doi.org/10.1016/j.indic.2019.100011

49. Nasseri A (2019) Energy use and economic analysis for wheat production by conservation tillage along with sprinkler irrigation. Sci Total Environ 648:450–459. https://doi.org/10.1016/j.scitotenv.2018.08.170

50. Notter B, Keller M, Althaus H, Cox B, Heidt C, Biemann K, Knorr W, Rader D, Jamet M (2020) HBEFA, 2014. Handb Emiss Factors Road Transp (HBEFA). Version HBEFA 4.1. https://www.hbefa.net/e/index.html

51. Ozkan B, Akcaoz H, Fert C (2004) Energy input-output analysis in Turkish agriculture. Renew Energy 29:39–51. https://doi.org/10.1016/S0960-1481(03)00135-6

52. Pandey D, Agrawal M, Bohra JS (2012) Greenhouse gas emissions from rice crop with different tillage permutations in rice-wheat system. Agric Ecosyst Environ. https://doi.org/10.1016/j.agee.2012.07.008

53. Pathak H, Aggarwal PK (2012) Low carbon technologies for agriculture: a study on rice and wheat systems in the indo-gangetic plains. New Delhi, India

54. Pathak H, Jain N, Bhatia A, Patel J, Aggarwal PK (2010) Carbon footprints of Indian food items. Agric Ecosyst Environ 139:66–73. https://doi.org/10.1016/j.agee.2010.07.002

55. Phyllis (2020) Database for biomass and waste. Wheat straw 1995. https://ukerc.rl.ac.uk/DC/cgi-bin/edc_search.pl/?WantComp=85

56. Pimentel D, Cooperstein S, Randell H, Filiberto D, Sorrentino S, Kaye B, Nicklin C, Yagi J, Brian J, O'Hern J, Habas A, Weinstein C (2007) Ecology of increasing diseases: population growth and environmental degradation. Hum Ecol 35:653–668. https://doi.org/10.1007/s10745-007-9128-3

57. Pishgar-Komleh SH, Ghahderijani M, Sefeedpari P (2012) Energy consumption and CO_2 emissions analysis of potato production based on different farm size levels in Iran. J Clean Prod 33:183–191. https://doi.org/10.1016/j.jclepro.2012.04.008

58. Rahman MM, Aravindakshan S, Hoque MA, Rahman MA, Gulandaz MA, Rahman J, Islam MT (2021) Conservation tillage (CT) for climate-smart sustainable intensification: assessing the impact of CT on soil organic carbon accumulation, greenhouse gas emission and water footprint of wheat cultivation in Bangladesh. Environ Sustain Indic 10:100106. https://doi.org/10.1016/j.indic.2021.100106

59. Rahman S, Hasan MK (2014) Energy productivity and efficiency of wheat farming in Bangladesh. Energy 66:107–114. https://doi.org/10.1016/j.energy.2013.12.070

60. Sayed A, Sarker A, Kim JE, Rahman MM, Mahmud MGA (2020) Environmental sustainability and water productivity on conservation tillage of irrigated maize in red brown terrace soil of Bangladesh. J Saudi Soc Agric Sci 19:276–284. https://doi.org/10.1016/j.jssas.2019.03.002

61. Singh BP, Setia R, Wiesmeier M, Kunhikrishnan A (2018) Agricultural management practices and soil organic carbon storage. Soil Carbon Storage: Modulators, Mech Model. https://doi.org/10.1016/B978-0-12-812766-7.00007-X

62. Singh G, Singh S, Singh J (2004) Optimization of energy inputs for wheat crop in Punjab. Energy Convers Manag 45:453–465. https://doi.org/10.1016/S0196-8904(03)00155-9

63. Singh H, Singh AK, Kushwaha HL, Singh A (2007) Energy consumption pattern of wheat production in India. Energy 32:1848–1854. https://doi.org/10.1016/j.energy.2007.03.001

64. Singh S, Singh S, Mittal JP, Pannu CJS (1998) Frontier energy use for the cultivation of wheat crop in punjab. Energy Convers Manag 39:485–491. https://doi.org/10.1016/s0196-8904(96)00234-8

65. Stocker TF, Qin D, Plattner G-K, Tignor M, Allen SK, Boschung J, Nauels A, Xia Y, Bex V, Midgley PM (2013) IPCC, 2013: climate change 2013: the physical science basis. In: Contribution of working group i to the fifth assessment report of the intergovernmental panel on climate change, IPCC

66. Suh S (2004) Materials and energy flows in industry and ecosystem networks. Int J Life Cycle Assess 9:335–336. https://doi.org/10.1007/bf02979425
67. UNEP (2017) The emissions gap report 2017—a UN environment synthesis report. United Nations Environ Programme (UNEP). https://doi.org/10.1016/j.biocon.2006.04.034
68. Volanti M, Martínez CC, Cespi D, Lopez-Baeza E, Vassura I, Passarini F (2021) Environmental sustainability assessment of organic vineyard practices from a life cycle perspective. Int J Environ Sci Technol. https://doi.org/10.1007/s13762-021-03688-2
69. Wang W, Yuan J, Gao S, Li T, Li Y, Vinay N, Mo F, Liao Y, Wen X (2020) Conservation tillage enhances crop productivity and decreases soil nitrogen losses in a rainfed agroecosystem of the Loess Plateau, China. J Clean Prod. https://doi.org/10.1016/j.jclepro.2020.122854
70. Wang Z, Wang G, Han Y, Feng L, Fan Z, Lei Y, Yang B, Li X, Xiong S, Xing F, Xin M, Du W, Li C, Li Y (2020) Improving cropping systems reduces the carbon footprints of wheat-cotton production under different soil fertility levels. Arch Agron Soil Sci. https://doi.org/10.1080/03650340.2020.1720912
71. Wiedmann T, Minx J (2007) A definition of ' carbon footprint. Science 80-:1–7. https://doi.org/10.1088/978-0-750-31040-6
72. Worldometer (2020) Bangladesh population. https://www.worldometers.info/world-population/bangladesh-population/. Accessed 22 Oct 21
73. Xiao L, Zhao R, Zhang X (2020) Crop cleaner production improvement potential under conservation agriculture in China: a meta-analysis. J Clean Prod. https://doi.org/10.1016/j.jclepro.2020.122262
74. Yousefi M, Damghani AM, Khoramivafa M (2014) Energy consumption, greenhouse gas emissions and assessment of sustainability index in corn agroecosystems of Iran. Sci Total Environ 493:330–335. https://doi.org/10.1016/j.scitotenv.2014.06.004
75. Yousefi M, Mahdavi Damghani A, Khoramivafa M (2016) Comparison greenhouse gas (GHG) emissions and global warming potential (GWP) effect of energy use in different wheat agroecosystems in Iran. Environ Sci Pollut Res 23:7390–7397. https://doi.org/10.1007/s11356-015-5964-7
76. Zangeneh M, Omid M, Akram A (2010) A comparative study on energy use and cost analysis of potato production under different farming technologies in Hamadan province of Iran. Energy 35:2927–2933. https://doi.org/10.1016/j.energy.2010.03.024
77. Zentner RP, Lafond GP, Derksen DA, Nagy CN, Wall DD, May WE (2004) Effects of tillage method and crop rotation on non-renewable energy use efficiency for a thin Black Chernozem in the Canadian Prairies. Soil Tillage Res. 77:125–136. https://doi.org/10.1016/j.still.2003.11.002
78. Zhang D, Shen J, Zhang F, Li Y, Zhang W (2017) Carbon footprint of grain production in China. Sci Rep. https://doi.org/10.1038/s41598-017-04182-x
79. Zhang W, Zheng C, Song Z, Deng A, He Z (2015) Farming systems in China: innovations for sustainable crop production. In: Crop physiology: applications for genetic improvement and agronomy, 2nd edn. https://doi.org/10.1016/B978-0-12-417104-6.00003-0
80. Ziaei SM, Mazloumzadeh SM, Jabbary M (2015) A comparison of energy use and productivity of wheat and barley (case study). J Saudi Soc Agric Sci 14:19–25. https://doi.org/10.1016/j.jssas.2013.04.002

Printed in the United States
by Baker & Taylor Publisher Services